D0860593

EVALUATION AND CONTROL OF MEASUREMENTS

QUALITY AND RELIABILITY

A Series Edited by

EDWARD G. SCHILLING

Coordinating Editor

Center for Quality and Applied Statistics
Rochester Institute of Technology
Rochester, New York

W. GROVER BARNARD

Associate Editor for Human Factors

Vita Mix Corporation
Cleveland, Ohio

RICHARD S. BINGHAM, JR.

Associate Editor for Quality Management

Consultant
Brooksville, Florida

LARRY RABINOWITZ

Associate Editor for Statistical Methods

College of William and Mary
Williamsburg, Virginia

THOMAS WITT

Associate Editor for Statistical Quality Control

Rochester Institute of Technology
Rochester, New York

1. Designing for Minimal Maintenance Expense: The Practical Application of Reliability and Maintainability, *Marvin A. Moss*
2. Quality Control for Profit, Second Edition, Revised and Expanded, *Ronald H. Lester, Norbert L. Enrick, and Harry E. Mottley, Jr.*
3. QCPAC: Statistical Quality Control on the IBM PC, *Steven M. Zimmerman and Leo M. Conrad*
4. Quality by Experimental Design, *Thomas B. Barker*
5. Applications of Quality Control in the Service Industry, *A. C. Rosander*
6. Integrated Product Testing and Evaluating: A Systems Approach to Improve Reliability and Quality, Revised Edition, *Harold L. Gilmore and Herbert C. Schwartz*

7. Quality Management Handbook, *edited by Loren Walsh, Ralph Wurster, and Raymond J. Kimber*
8. Statistical Process Control: A Guide for Implementation, *Roger W. Berger and Thomas Hart*
9. Quality Circles: Selected Readings, *edited by Roger W. Berger and David L. Shores*
10. Quality and Productivity for Bankers and Financial Managers, *William J. Latzko*
11. Poor-Quality Cost, *H. James Harrington*
12. Human Resources Management, *edited by Jill P. Kern, John J. Riley, and Louis N. Jones*
13. The Good and the Bad News About Quality, *Edward M. Schrock and Henry L. Lefevre*
14. Engineering Design for Producibility and Reliability, *John W. Priest*
15. Statistical Process Control in Automated Manufacturing, *J. Bert Keats and Norma Faris Hubele*
16. Automated Inspection and Quality Assurance, *Stanley L. Robinson and Richard K. Miller*
17. Defect Prevention: Use of Simple Statistical Tools, *Victor E. Kane*
18. Defect Prevention: Use of Simple Statistical Tools, Solutions Manual, *Victor E. Kane*
19. Purchasing and Quality, *Max McRobb*
20. Specification Writing and Management, *Max McRobb*
21. Quality Function Deployment: A Practitioner's Approach, *James L. Bossert*
22. The Quality Promise, *Lester Jay Wollschlaeger*
23. Statistical Process Control in Manufacturing, *edited by J. Bert Keats and Douglas C. Montgomery*
24. Total Manufacturing Assurance, *Douglas C. Brauer and John Cesarone*
25. Deming's 14 Points Applied to Services, *A. C. Rosander*
26. Evaluation and Control of Measurements, *John Mandel*

EVALUATION AND CONTROL OF MEASUREMENTS

John Mandel
Silver Spring, Maryland

Marcel Dekker, Inc. **New York • Basel • Hong Kong**

Library of Congress Cataloging--in--Publication Data

Mandel, John
Evaluation and control of measurements/John Mandel.
p. cm. -- -- (Quality and reliability; v. 26)
Includes bibliographical references and index.
ISBN 0-8247-8531-2
1. Mathematical statistics. I. Title. II. Series.
QA276.M326 1991
519.5-- --dc20 91-14138
CIP

This book is printed on acid-free paper.

Copyright © 1991 by MARCEL DEKKER, INC. All Rights Reserved

Neither this book nor any part may be reproduced or transmitted in any form or by any means, electronic or mechanical, including photocopying, microfilming, and recording, or by any information storage and retrieval system, without permission in writing from the publisher.

MARCEL DEKKER, INC.
270 Madison Avenue, New York, New York 10016

Current printing (last digit):
10 9 8 7 6 5 4 3 2 1

PRINTED IN THE UNITED STATES OF AMERICA

About the Series

The genesis of modern methods of quality and reliability will be found in a simple memo dated May 16, 1924, in which Walter A. Shewhart proposed the control chart for the analysis of inspection data. This led to a broadening of the concept of inspection from emphasis on detection and correction of defective material to control of quality through analysis and prevention of quality problems. Subsequent concern for product performance in the hands of the user stimulated development of the systems and techniques of reliability. Emphasis on the consumer as the ultimate judge of quality serves as the catalyst to bring about the integration of the methodology of quality with that of reliability. Thus, the innovations that came out of the control chart spawned a philosophy of control of quality and reliability that has come to include not only the methodology of the statistical sciences and engineering, but also the use of appropriate management methods together with various motivational procedures in a concerted effort dedicated to quality improvement.

This series is intended to provide a vehicle to foster interaction of the elements of the modern approach to quality, including statistical applications, quality and reliability engineering, management, and motivational aspects. It is a forum in which the subject matter of these various areas can be brought together to allow for effective integration of appropriate techniques. This will promote the true benefit of each, which can be achieved only through their interaction. In this sense, the whole of quality and reliability is greater than the sum of its parts, as each element augments the others.

The contributors to this series have been encouraged to discuss fundamental concepts as well as methodology, technology, and procedures at the leading edge of the discipline. Thus, new concepts are placed in proper perspective in these evolving disciplines. The series is intended for those in manufacturing, engineering, and marketing and management, as well as the consuming public, all of whom have an interest and stake in the improvement and maintenance of quality and reliability in the products and services that are the lifeblood of the economic system.

The modern approach to quality and reliability concerns excellence: excellence when the product is designed, excellence when the product is made, excellence as the product is used, and excellence throughout its lifetime. But excellence does not result without effort, and products and services of superior quality and reliability require an appropriate combination of statistical, engineering, management, and motivational effort. This effort can be directed for maximum benefit only in light of timely knowledge of approaches and methods that have been developed and are available in these areas of expertise. Within the volumes of this series, the reader will find the means to create, control, correct, and improve quality and reliability in ways that are cost effective, that enhance productivity, and that create a motivational atmosphere that is harmonious and constructive. It is dedicated to that end and to the readers whose study of quality and reliability will lead to greater understanding of their products, their processes, their workplaces, and themselves.

Edward G. Schilling

Preface

This book covers a number of topics dealing with the analysis of data. It is also concerned with the quality control of measurements. Experimental data come in such a variety of forms that a comprehensive treatment is impossible. Nevertheless, I discuss general principles that should make it possible to handle many different situations. After reading this book, an experimental scientist should be able to develop the tables and the graphs that will reveal to him the general nature as well as the fine points of his data. He need not be a mathematician or a statistician to understand this book: a knowledge of high-school algebra is sufficient for understanding practically everything discussed in it. In only a few places is elementary calculus used.

The book is unorthodox. I do not apologize for this for I believe that data analysis is in general not aided by applying established statistical techniques such as analysis of variance and covariance. This is so because data sets very seldom follow the models underlying these techniques. In fact, they seldom conform to any model at all, because different subgroups of the data often exhibit patterns that are not shared by other subgroups. The pooling inherent in analysis of variance and in analysis of covariance often hides real differences, which must be discovered by the data analyst, often by trial and error. Thus, the analysis of real data sets is to a large extent governed by the inventiveness of the analyst. This does not mean that analysts can dispense with statistical knowledge. Indeed, in addition to using intuition, they should make extensive use of the basic facts of statistical theory. Therefore, the first chapters of this book deal with the fundamentals of statistics. In the following chapters many common situations are explored, almost always in terms of real data, to illustrate the use of statistical thinking in data analysis. Each of these chapters also introduces new techniques appropriate to its subject.

This book should be useful not only to the experimental scientist, but also to the statistical consultant, who will find in it a new approach to many problems. It is hoped that this book will contribute to a better way of exploring and understanding scientific data.

John Mandel

Contents

EVALUATION AND CONTROL OF MEASUREMENTS

1

Measurement and Statistics

1.1 MEASUREMENT AS A PROCESS

Manufactured products are the end product of a process called a manufacturing process. Similarly, measurements are the end product of a process that we call a measuring process. A manufacturing process involves equipment that must be set up, adjusted, and monitored during the manufacturing process. Analogously, a measuring process involves instrumentation that must be calibrated and kept in a state of proper calibration throughout the process of measuring. In a manufacturing process, the elements going into the process are the raw materials; the analog to the raw materials in a measuring process are the chemical products or physical objects that are to be measured, such as a sample of blood, to be measured for its glucose content, or a microscope lens, to be measured for its optical properties.

From the analogy of manufacturing and measuring processes, it follows that there exists a need for the *quality control* of measurements, just as there is a need for controlling the quality of a product. Indeed, if we purchase a new carburetor for our automobile, we expect that it will be a bona fide member of an entire population of carburetors of the indicated type, characterized by the fact that they are all reliable replacements for each other. Similarly, measurements of glucose made on the same sample of blood serum in different laboratories should be, if not totally identical, at least close enough to each other, so that all practical conclusions (e.g., diagnostics) based on them will be the same. To attain this goal it is not sufficient to install in each laboratory good instrumentation, purchased from a reliable source; it is also necessary to maintain the measuring instrumentation, and the way it is used by the laboratory personnel, in a state of control.

The need for assuring the quality control of manufactured products has been universally recognized and is in fact the object of numerous books, courses, and seminars presently popular in this and other countries. Unfortunately, the same cannot be said of measuring processes. This is ironic in that quality control procedures of products always involve measurements and are in fact based on the measurements made to monitor the quality of the product and the proper functioning of the manufacturing equipment. Thus, quality control of products is impossible without the availability of reliable measuring processes.

The statistical methods developed for the control of manufactured goods are also applicable to the control of measuring equipment. One needs one or more homogeneous materials, samples of which are submitted periodically to the measuring process that is to be monitored. This may apply either to single laboratories or, in the case of interlaboratory comparison studies, to a number of participating laboratories who had agreed to join in the study. We will see that such studies require careful planning as well as serious study of the data obtained from them.

1.2 MEASUREMENT AS A RELATION

The view of measurement as a process is particularly useful in the context of the quality control of measurements. It does not, by itself, reveal all the properties of measurement. Some of these become apparent only when the process is studied on an interlaboratory scale. Thus we are led to look at measurement from a second viewpoint: measurement as a relation.

As a rule, measurements are indirect evaluations of properties of materials or systems. Thus, when determining the lead content in a sample of paint, we do not count the lead atoms in a given mass of paint, but transform the lead, through a chemical reaction, into a compound that can be isolated, if necessary, and of which a given physical or chemical characteristic can be directly measured. In order to obtain the desired answer, in terms of lead, we then need the relationship between lead and this measured characteristic. Such a relation can be obtained either from theoretical considerations or from a calibration experiment.

Mathematically, we visualize a rectangular plot in which the desired property, say P, is represented by the x-axis, and the measured characteristic, say M, by the y-axis. The relation is then represented by a monotonic curve of M versus P. Monotonic means, either increasing or decreasing over the entire range, but not both. This is necessary because nonmonotonicity would result in more than one value of M for a given value of P and thus would destroy the usefulness of M as a measure of P.

The calibration experiment for the establishment of the P,M relation can be carried out by measuring M for a series of samples of different P values, covering the range of P in which one is interested. This requires knowledge of P for these samples, obtained either from theory, or by synthesizing these samples with known amounts of P, or by measuring them by means of a different, well-established measuring technique.

Table 1.1 presents an example of calibration data of the type we discussed. Here P is the furfural concentration of samples of wood pulp and M is "absorbency," a colorimetric result obtained on a distillate of the pulp, by means of a method known as the "orcinol" method. The chemical details can be found in the paper from which these data are taken (Wilson and Mandel, 1960) and are not required for our discussion. A plot of the data is shown in Fig. 1.1. The calibration curve is clearly not linear, a fact we mention to alert the

Table 1.1 Calibration Data for Determination of Furfural by the Orcinol Method

Furfural concentration (mg/l)	Mean absorbency
50	0.109
100	0.211
200	0.429
350	0.792
500	1.185

reader to the frequent occurrence of nonlinear calibration curves. Actually, the calibration curve in our example is a composite of 19 calibration experiments over a period of 14 months, reported by a Swedish scientist. For a more detailed discussion of measurement as a relation, the primary question concerns the reproducibility of this calibration curve, within a laboratory and between laboratories. Such a study requires a different experiment, known as an interlaboratory test, or round robin. It is merely an extension of the calibration experiment to a number of laboratories, taking care, however, to send samples of the same set of materials to all participant laboratories. Such a study was carried out for the orcinol method and is reported in the same paper (Wilson and Mandel, 1960). Its results are reported however, not in terms of *furfural,* but rather as concentrations of *pentosans,* a result obtained by a simple conversion formula. The results of the interlaboratory study are shown in Table 1.2. We will discuss this table and its analysis in detail in a later chapter. At this point we merely call attention to two items. First, it is seen that each laboratory made three determinations on each of the nine materials. This replication is common in interlaboratory tests, for reasons to be explained when we discuss this subject in a general way. Secondly, it is apparent, upon examination of the data, that if calibration curves were drawn separately for each laboratory, the nine calibration curves would not all be identical. In general, the calibration curves corresponding to the laboratories in an interlaboratory test show systematic

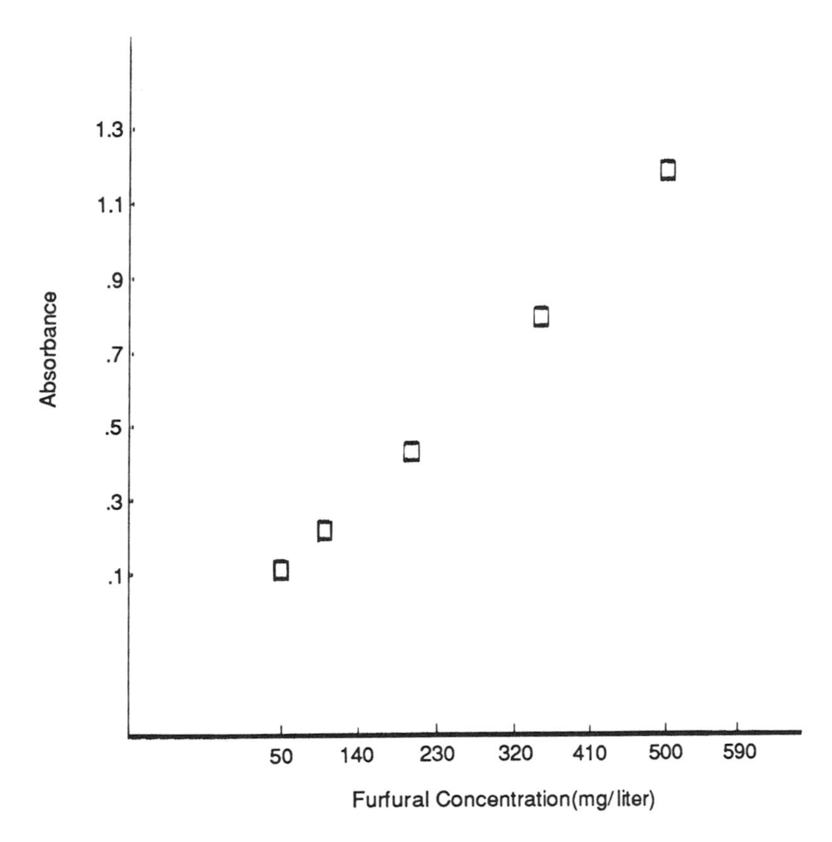

Figure 1.1 Determination of furfural by orcinol method.

differences from each other. The nature and magnitude of these systematic differences are important features of the test method that is being subjected to the interlaboratory study.

1.3 STATISTICAL POPULATIONS AND SAMPLES

Populations and samples are fundamental concepts of statistics, just as atoms, molecules, force, and energy are fundamental concepts in the physical sciences. Like these concepts, they are constructs of the mind expressible in terms of mathematical models, rather than concrete items.

Returning to our discussion of measurement as a process, let us

Table 1.2 Pentosans in Pulp

Lab / *Material*	*A*	*B*	*C*	*D*	*E*	*F*	*G*	*H*	*I*
	0.44	0.96	1.23	1.25	1.98	4.12	5.94	10.70	17.13
1	0.49	0.92	1.88	1.25	1.92	4.16	5.37	10.74	16.56
	0.44	0.82	1.24	1.42	1.80	4.16	5.37	10.83	16.56
	0.41	0.83	1.12	1.25	1.99	4.10	5.26	10.07	16.08
2	0.41	0.83	1.12	1.25	1.94	4.11	5.26	10.05	16.04
	0.41	0.84	1.12	1.26	1.95	4.10	5.26	9.82	16.13
	0.51	0.92	1.11	1.35	2.05	4.11	5.16	10.01	16.01
3	0.51	0.93	1.13	1.35	2.08	4.16	5.16	10.17	15.96
	0.51	0.92	1.11	1.35	2.03	4.16	5.21	10.17	16.06
	0.40	0.96	1.15	1.29	2.05	4.20	5.20	10.98	16.65
4	0.38	0.94	1.13	1.29	2.04	4.20	5.20	10.67	16.91
	0.37	0.94	1.13	1.29	2.04	4.22	5.20	10.52	16.75
	0.49	0.82	0.98	1.23	1.94	4.61	5.00	10.48	15.71
5	0.49	0.82	0.98	1.23	1.96	4.63	5.00	10.27	15.45
	0.49	0.84	0.98	1.23	1.96	4.53	4.96	10.38	15.66
	0.43	0.88	1.11	1.31	2.01	3.93	4.85	9.57	15.05
6	0.41	0.92	1.12	1.30	1.99	3.92	4.87	9.57	14.73
	0.40	0.88	1.11	1.31	1.98	3.84	4.91	9.62	15.04
	0.186	0.866	1.05	1.13	1.98	4.21	5.27	11.5	18.8
7	0.171	0.900	0.962	1.15	1.93	4.18	5.32	10.8	18.2
	0.153	0.831	0.927	1.15	1.98	4.16	5.10	11.5	18.1

imagine that samples from a single, homogeneous source are fed consecutively to a specified measuring process. To fix ideas, consider a large supply of serum, in which the concentration of cholesterol has to be determined. The serum, after collection, is thoroughly mixed and successive portions are analyzed for cholesterol. Conceptually, we can visualize a very large succession of portions, each giving rise to a single result expressed as milligrams per deciliter of cholesterol. In fact, since this is only a thought experiment, we can visualize an unending series of portions, and therefore also of numerical measurements, of cholesterol content. Such a sequence has been called an hypothetical infinite population (Fisher, 1941). The numerical values constituting this population will not all be identical, because each measurement is affected by experimental error. The presence of experimental error is due to unavoidable fluctuations in the conditions (temperature, pressure, concentration of reagents, humidity, etc.) during the measuring process. Thus, the fundamental characteristic of a statistical population is *variability*. In a process that is *in statistical control,* the fluctuations causing the variability in the results are of *a random nature,* and so are the measurements themselves. We will not try to define randomness, but rather rely on the reader's intuition for understanding this concept. Of course, a basic feature of randomness is *unpredictability,* except in a statistical sense. Thus, having looked at, say, the first ten measurements, we cannot predict where exactly, in the range of possible values, the eleventh measurement will lie. We could, however, make predictions in terms of *probabilities,* as we will learn shortly.

In practice one never obtains a population of values, simply because by definition of the hypothetical infinite population, the number of members of this population is infinite. What one *does* obtain is a *statistical sample,* that is, a finite collection of items from the population, for example, the first ten measurements. We would call this a sample *of size ten*. Figure 1.2 is a schematic representation of a statistical sample. The values, representing measurements of the cholesterol content of homogeneous source of serum, are plotted on a vertical scale, against the ordinal values 1 to 10 plotted on horizontal scale. Of course, the sequence of measurements could, and in fact, has been continued. The abstract concept of a statistical population is now readily grasped, by imagining that this sequence is continued indefinitely. In Table 1.3 and in Fig. 1.3, the sequence has been continued to 25 measurements. Furthermore, a *frequency diagram* has been constructed in Fig. 1.3 by displacing each value horizontally to the right and lining the values up against a vertical line. This operation reveals a new feature of the data: the values in the middle of the range

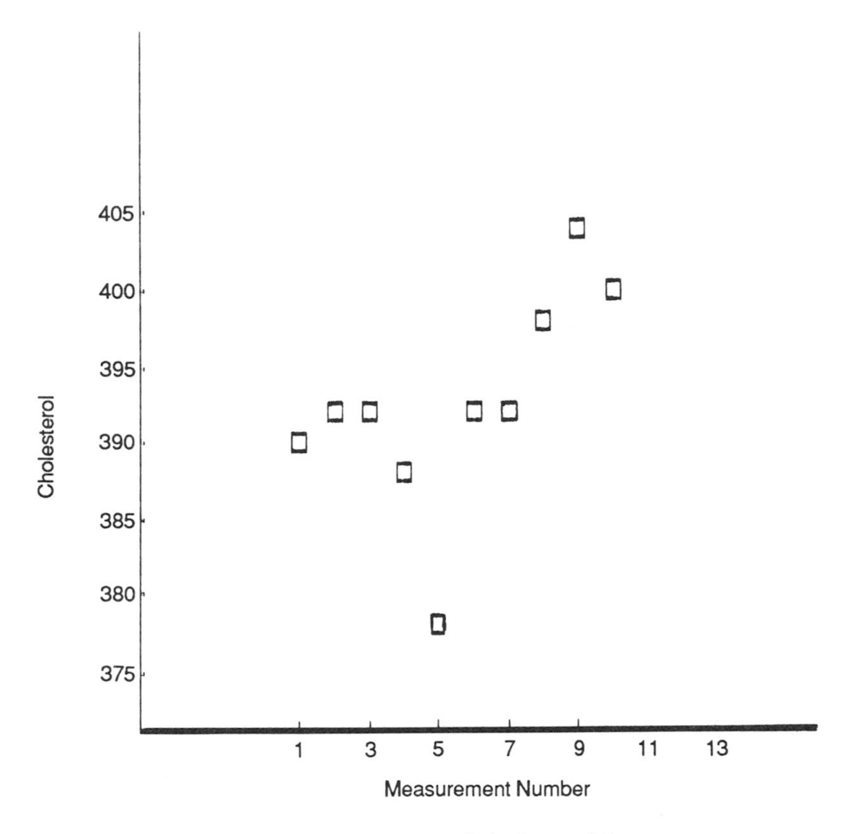

Figure 1.2 First ten measurements of cholesterol in serum.

are more frequent than those on the high side or on the low side. The *frequency diagram* displays the different frequencies with which measurements of different magnitudes occur. It contains however other information as well: by labeling the vertical axis on the right, we see that the central value of the sample of 25 values is around 394 and that the total width covered by them is 378 to 404. More satisfactory measures of these two quantities are obtained as follows.

1. A measure of the central value is given by the *mean,* which is simply the arithmetic average of all values. If the original measurements are denoted by x_1, x_2, ..., x_{25}, then the mean is denoted by the symbol $\bar{x}$.

Table 1.3 A Sample of 25 Cholesterol Measurements

Observation Number (i)	Measured value (x_i)	Deviation (d_i)	Deviation squared (d_i^2)
1	390	−1.92	3.6864
2	392	0.08	0.0064
3	392	0.08	0.0064
4	388	−3.92	15.3664
5	378	−13.92	193.7664
6	392	0.08	0.0064
7	392	0.08	0.0064
8	398	6.08	36.9664
9	404	12.08	145.9264
10	400	8.08	65.2864
11	402	10.08	101.6064
12	392	0.08	0.0064
13	398	6.08	36.9664
14	380	−11.92	142.0864
15	398	6.08	36.9664
16	388	−3.92	15.3664
17	402	10.08	101.6064
18	386	−5.92	35.0464
19	386	−5.92	35.0464
20	390	−1.92	3.6864
21	396	4.08	16.6464
22	396	4.08	16.6464
23	384	−7.92	62.7264
24	388	−3.92	15.3664
25	386	−5.92	35.0464
	$\bar{x}$ = 391.92	sum = 0	sum = 1115.84

$$s^2 = \frac{\Sigma d_i^2}{25 - 1} = \frac{1115.84}{24} = 46.4933,\ s = \sqrt{46.4933} = 6.8186.$$

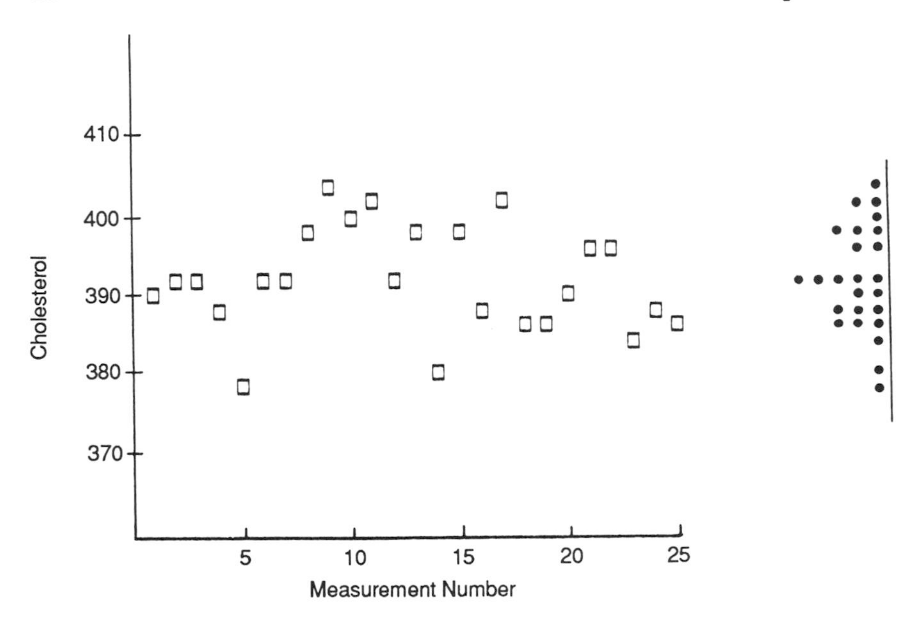

Figure 1.3 Twenty-five measurements of cholesterol in serum.

$$\bar{x} = \frac{x_1 + x_2 + \cdots + x_{25}}{25} = \frac{\sum (x)}{N} \tag{1.1}$$

where Σ denotes summation and N is the size of the sample.

2. A measure of width is given by the *standard deviation,* denoted by s and given by the formula

$$s = \sqrt{\frac{\sum (x_i - \bar{x})^2}{N - 1}} \tag{1.2}$$

Another related measure of width is the *variance,* which is simply the square of the standard deviation. It is denoted by the symbol s^2:

$$s^2 = \frac{\sum (x_i - \bar{x})^2}{N - 1} \tag{1.3}$$

Let us denote the "deviation" of each observation from the mean by the symbol d; then we have

$$d_i = x_i - \bar{x} \tag{1.4}$$

$$s^2 = \frac{\sum d_i^2}{N - 1} \tag{1.5}$$

and

$$s = \sqrt{\sum \frac{d_i^2}{N - 1}} \tag{1.6}$$

The value $N - 1$ used in the denominators of Eqs. (1.3, 1.5, and 1.6) requires an explanation. If instead of $N - 1$ we had written N in Eq. (1.3), then s^2 would simply be the average of the squares of the deviations. The reason for using $N - 1$ rather than N is that the N deviations are not independent. This is readily seen from Table 1.3, which also shows the computation of the mean and the standard deviation. It is readily verified that the sum of the deviations is zero and that this would be true, no matter what the values of the original x_i are, it being a mathematical consequence of the definition of the mean. Thus, given 24 of the 25 deviations, the 25th is entirely determined by the other 24. Therefore, only $N - 1$ of the N deviations are "free to vary." We indicate this by saying that the deviations have $N - 1$ "degrees of freedom" and use the degrees of freedom in the denominator of the measures of width, rather than the sample size N.

The standard deviation and the variance are often referred to as measures of *dispersion*.

The sum of squares of deviations, Σd_i^2, is often referred to as simply the "sum of squares" and denoted by the symbol SS. Furthermore, the degrees of freedom are often represented by the symbol DF. Thus we have:

$$s^2 = \frac{\text{SS}}{\text{DF}} \quad \text{and} \quad s = \sqrt{\frac{\text{SS}}{\text{DF}}} \tag{1.7}$$

It is also common to refer to the ratio SS/DF in a general way as the "mean square," denoted by the symbol MS. Hence

$$\mathrm{MS} = \frac{\mathrm{SS}}{\mathrm{DF}} \tag{1.8}$$

It is useful to observe that the total range of values in the example, 404 − 378 = 26, is about 4 standard deviations. For sample sizes N up to 15, an approximate formula for the ratio of the range to the standard deviation is given by

$$\frac{w}{s} \cong \sqrt{N} \tag{1.9}$$

where w denotes the range (largest value minus smallest value) of the sample. However, this formula should not be used for sample sizes N larger than 15. For large N, one will find values of w/s of the order of 6.

The measuring process for the determination of cholesterol is of course not properly characterized by the frequency diagram shown in Fig. 1.3, nor by the values of the mean and standard deviation derived from these 25 values. This is evident from the fact that these values, as well as the diagram, would have been different, had we taken 25 more measurements and based our calculations on them. We can however visualize a much larger sample, say of 25,000 values rather than 25. Then the diagram and the values of x and s would have been better representations of the measuring process. By a conceptual generalization, we visualize a sample of very large ("infinite") size to define the frequency diagram of the measuring process. We then call it the "frequency distribution." The values of x and s obtained for this very large sample we then call the *population mean* and the *population standard deviation* and we represent them by the Greek letters μ and σ. We call these values *population parameters,* whereas x and s obtained from a finite sample are called *sample estimates* of the corresponding parameters. The larger the sample, the more likely it becomes that $\bar{x}$ will be close to μ, and s close to σ. We will see that this statement can be cast in a more quantitative form.

In Table 1.4 we show $\bar{x}$ and s values for the five samples of size 5 representing, respectively, measurements 1 through 5, 6 through 10, 11 through 15, 16 through 20, and 21 through 25 of Table 1.3. We

Table 1.4 Values of $\bar{x}$ and s for Subsets of Five of Table 1.3

Sample	Subset of measurements	$\bar{x}$	s
1	1 through 5	388.0	5.8310
2	6 through 10	397.2	5.2154
3	11 through 15	394.0	8.6023
4	16 through 20	390.4	6.6933
5	21 through 25	390.0	5.6569

call attention to the large variability of these values or, in other words, to their rather severe lack of stability. It is clear that sample estimates of population parameters, especially when they are based on small samples (say $N < 20$), are quantities subject to the interplay of chance effects. We will say that they are "subject to sampling fluctuations." By contrast, μ and σ are constants characteristic of the entire population of values, a population we identify in some sense with the measuring process in question.

1.4 THE STATISTICAL APPROACH TO MEASUREMENT

We pause at this rather early stage in our development to emphasize a fundamental feature of the approach which this book attempts to represent.

A good experimental scientist will spare no efforts to assure that his measurements are as good as he can make them. Having made a measurement, he therefore will generally have implicit faith in it, as a valid measure of the property he intended to quantify. His interest is usually centered more in this property than in the characteristics of the measurement process he has used to measure it. The statistician, as data analyst, directs his attention to the measurements themselves, and to what he can infer from the observed measurements about the measuring process used to generate them. More specifically, the statistician regards each measurement as only one member of an infinite sequence of similar measurements, a "hypothetical infinite population," of which he seeks to establish the more important features. The reason for this is that any single measurement, while providing relevant information about the property it measures, provides no information on its own stability. This information can only be obtained from a study of the measuring process itself, through an analysis of a reasonably large number of measurements obtained by it.

Thus, the data analyst complements the work of the scientist by providing him with information about the quality of his measurements. It is then logical that he should also direct his attention to the way in which the measurements were made, and more generally to the *design of the experiment* from which the data were obtained. Furthermore, the statistician, or data analyst, should also be concerned with methods that will monitor the quality of the measurements over a period of time, to ensure that stability is obtained to the largest possible extent. We will attempt to represent in this book ways and means for achieving these aims.

REFERENCES

Fisher, R. A. (1941). Statistical Methods for Research Workers. G.E. Stechert & Co., New York.

Wilson, W. K, and J. Mandel (1960). Determination of Pentosans. Interlaboratory Comparison of the Aniline Acetate Orcinol and Bromination Methods. ACS-ATM-TAPPI-ICCA Pentosans Task Group. TAPPI, *43*.

2

Basic Statistical Concepts

2.1 INTRODUCTION

In the previous chapter we have mentioned the necessity to define and to use statistical terminology and notation to describe statistical properties. In this chapter we discuss the basic concepts of statistics as well as their interrelationships. A proper understanding of these concepts is essential for an intelligent use of statistics in the sciences.

2.2 FREQUENCY DISTRIBUTIONS

It is customary to present frequency diagrams, such as the one in Fig. 1.3, in a slightly different way, obtained by rotating it by an angle of 90 degrees clockwise from its position in Fig. 1.3. Then, the x-axis represents the measurement value and the y-axis the frequency with which this value occurs. Since a population is essentially a sample for

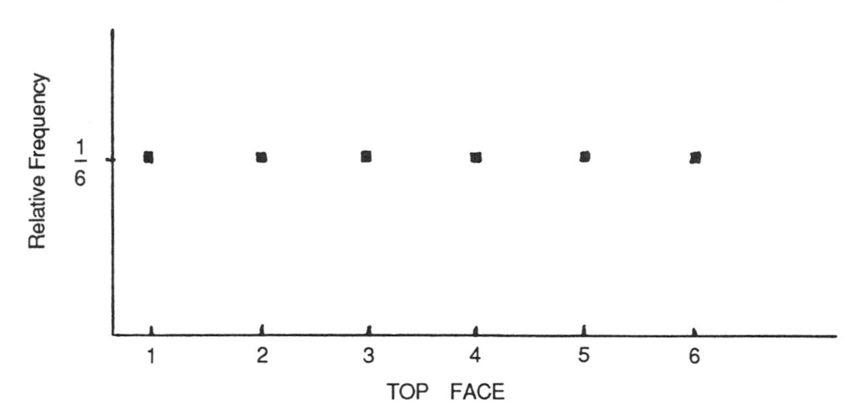

Figure 2.1 Probability in throwing a die.

which N approaches infinity, the frequencies will become very large as N increases in order to obtain the frequency distribution for the population. Therefore the frequencies are expressed as *relative frequencies,* defined as the ratios of the individual frequencies to N, the total sample size. The frequency distribution is then the limiting plot, for N tending to infinity, of relative frequencies versus magnitude of the measurement. The variable representing the value of the measurement is called a variate.

There are many different types of frequency distributions.

First we mention the *discrete* distributions, exemplified by Fig. 2.1 and 2.2. The first represents the relative frequencies of obtaining the values 1, 2, . . . , 6 on the top face when tossing a die. If the die is unbiased, each of these 6 possibilities has equal probability of occurring, and therefore a relative frequency of 1/6. Figure 2.2 is the frequency distribution for the sum of the values on the top faces of two dice. Here the 11 possible outcomes, 2, 3, . . . , 12, no longer have equal frequencies and the distribution is triangular in shape. The distributions shown in Fig. 2.1 and Fig. 2.2 are both *discrete,* in the sense that the variate can only take specific values such as 1, 2, . . . , 12. The opposite of discrete is *continuous,* a more common case for measurement data. In a continuous distribution, the variate varies in a continuous way from its lowest to its highest value. Often these limiting values are $\pm\infty$.

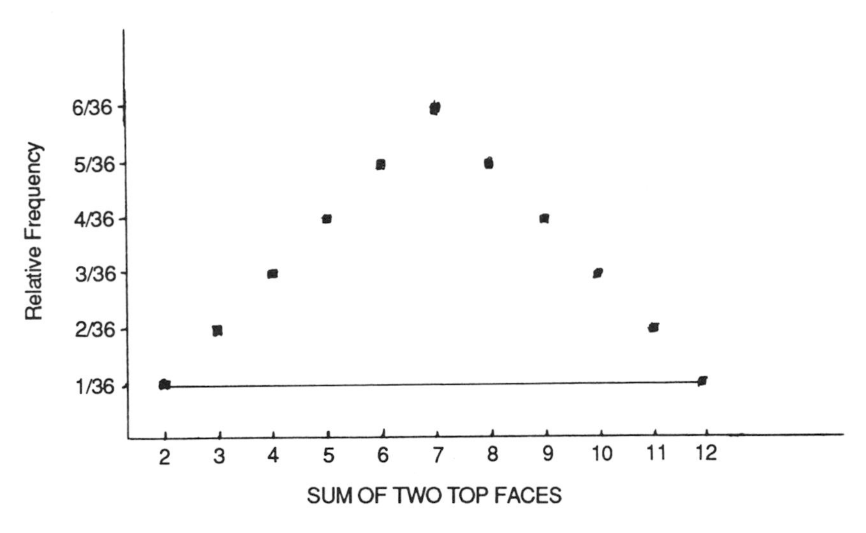

Figure 2.2 Probability in throwing two dice.

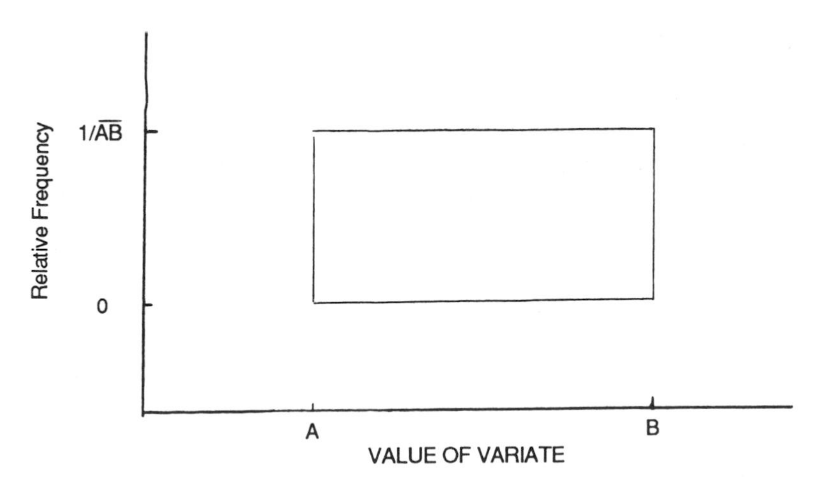

Figure 2.3 Rectangular distribution.

The simplest continuous distribution is the *rectangular*, shown in Fig. 2.3. Here the variate can have any real value between A and B, and the frequency of each of these values is the same, as shown by the horizontal line forming the top of the rectangle, hence the name rectangular. Figure 2.4 represents a more general continuous distribution.

It is easy to see that for *any* distribution the sum of all relative frequencies is *unity*. It then follows that for a continuous distribution, the relative frequency for any specific value of the variate is infinitesimally small. Consequently, for continuous distributions we introduce a different concept: the *frequency density*.

In Fig. 2.5, which represents a continuous distribution, consider a small interval such as CD. We consider the total frequency of all variate values between C and D. Intuition tells us that it is the ratio of the area of the shaded figure to the total area. The latter is however unity, since it is the sum of all possible relative frequencies. Hence, the relative frequency for the interval CD is the area of the pseudo-rectangle of which CD is the base and the frequency curve is the upper boundary.

Comparing the relative frequencies corresponding to intervals CD and EF we find that the latter is larger, in spite of the fact that CD and EF are intervals of equal length. We express this fact by stating that the *frequency density* is higher in the interval EF than it is in CD. More rigorously, we define frequency density as the *mathematical limit* of the ratio of the frequency corresponding to an interval (such as CD) to the length of the interval, as this length becomes smaller and smaller and finally tends to zero. In the limiting case, the interval has shrunk to a point on the variate axis and the frequency density just defined is simply the height of the curve at that point.

To make this concept more concrete, we will calculate the frequency density for a point of the rectangular distribution. Referring to Fig. 2.6, we consider the ratio of the area of the rectangle $CDD'C'$ to the base CD. The area in question is $CD \cdot h$, where h is the "height" of the distribution curve. Consequently, the frequency density is $CD \times h/CD = h$. But we can easily calculate h by observing that the area of the total extent of the frequency distribution is unity. Hence: $AB \times h = 1$, and consequently $h = 1/AB$. Thus, the frequency density for any point of the rectangular distribution is $1/AB$, and hence the

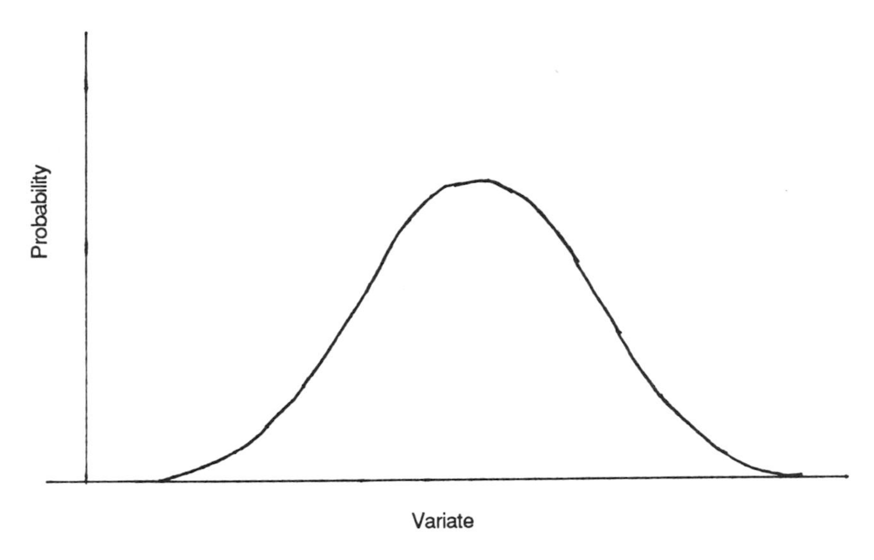

Figure 2.4 A continuous distribution.

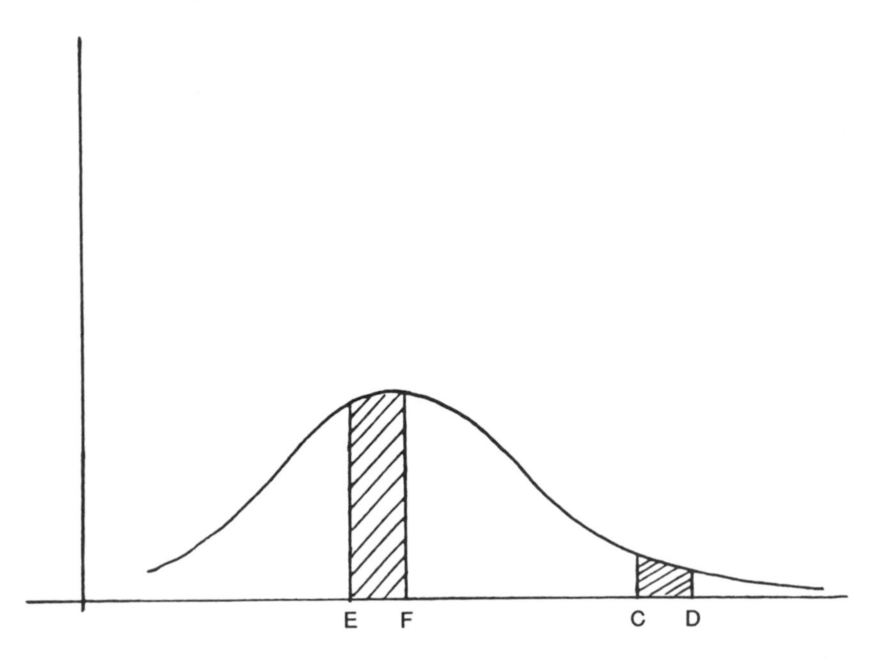

Figure 2.5 Concept of frequency density.

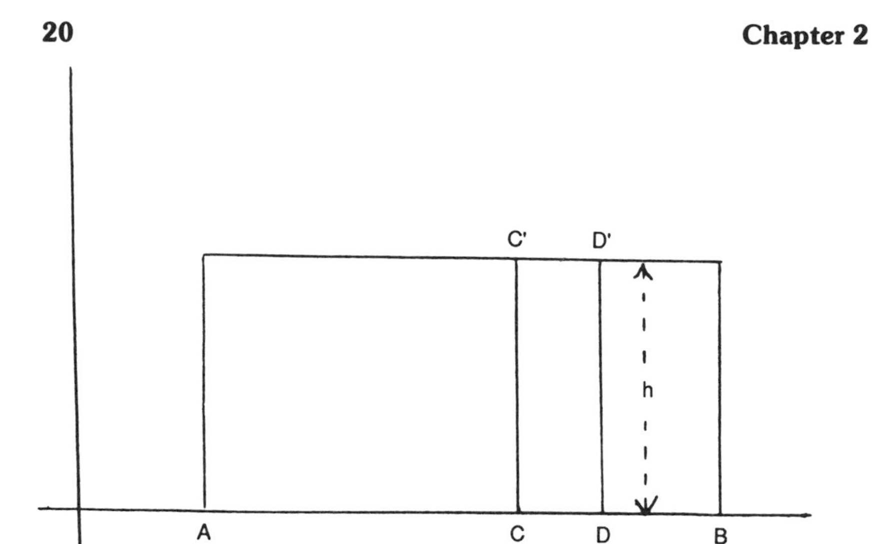

Figure 2.6 Frequency density in rectangular distribution.

same for all points. If, as is often done, we rescale our variate in the rectangular distribution in such a way that $A = 0$ and $B = 1$, then $AB = 1$ and the frequency density at any point of this distribution is unity.

As a second example, we consider the "triangular" distribution

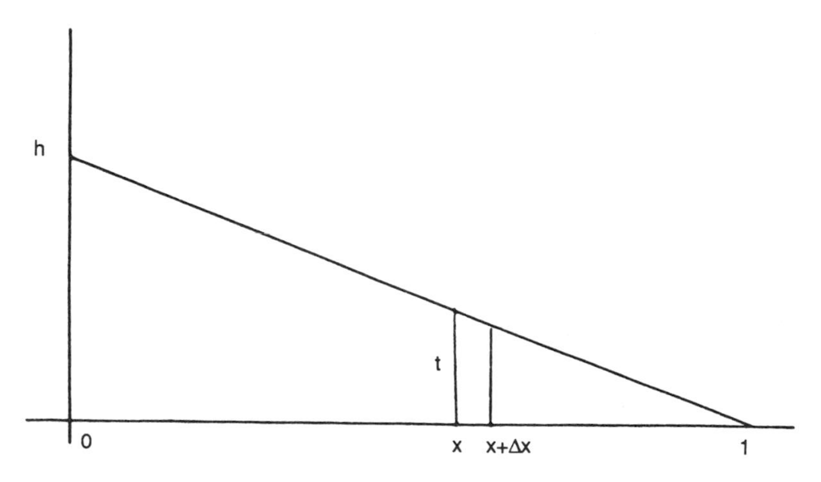

Figure 2.7 Frequency density in triangular distribution.

shown in Fig. 2.7, and we calculate the frequency density at point x. By definition, we have

$$\text{Frequency density} = \lim_{\Delta x \to 0} \frac{t \cdot \Delta x}{\Delta x} = t$$

From a consideration of similar triangles we have

$$\frac{t}{1 - x} = \frac{h}{1} \qquad \text{hence } t = h(1 - x)$$

On the other hand, since the total area is unity,

$$\frac{h \cdot 1}{2} = 1, \quad \text{hence } h = 2$$

Finally,

$$\text{Frequency density} = t = h(1 - x) = 2\,(1 - x)$$

2.3 WEIGHTED AVERAGES AND EXPECTED VALUES

Suppose that we have obtained the following set of measurements:

262, 281, 262, 274, 281, 280, 280, 262

The average is

$$\frac{262 + 281 + 262 + 274 + 281 + 280 + 280 + 262}{8}$$

$$= 272.75$$

If we note that some of the values occur more than once, we can write the numerator more expeditiously as

$$(3 \times 262) + (2 \times 281) + (2 \times 280) + 274$$

or

$$(3 \times 262) + (2 \times 281) + (2 \times 280) + (1 \times 274)$$

and the mean as

$$\frac{(3 \times 262) + (2 \times 281) + (2 \times 280) + (1 \times 274)}{3 + 2 + 2 + 1}$$

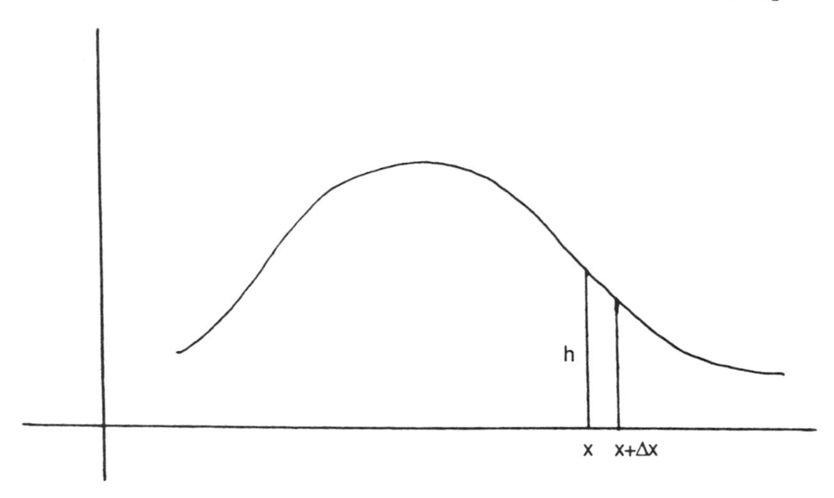

Figure 2.8 Frequency density in continuous distribution.

More generally, if any particular value x_i occurs n_i times, the formula for the mean becomes

$$\bar{x} = \frac{\Sigma n_i x_i}{\Sigma n_i} \tag{2.1}$$

This is called a *weighted average* or *weighted mean* where each value x_i is "weighted" by its frequency of occurrence. Equation (2.1) can also be written in the form:

$$\bar{x} = \sum\left(\frac{n_i}{\Sigma n_i}\right) x_i \tag{2.2}$$

which shows that the weighted mean is the sum of the products of each value by its relative frequency. The relative frequency $(n_i/\Sigma n_i)$ can be written more simply as f_i. Consider now a small interval, going from x to $x + \Delta x$, on the abscissa of a frequency distribution (see Fig. 2.8). Since the interval is small, we can equate each value x_i in it with its first value x. The relative weight for the entire interval is the area of the pseudorectangle erected over x, and is approximately equal to $\Delta x \cdot h$ where h is the height of the pseudorectangle. Thus, the weighted average for the interval $(x, x + \Delta x)$ is equal to $x \Delta x h$. But we have seen that h is the frequency density, f, at the point x. Thus, the weighted average for the interval is $x \cdot \Delta x \cdot f$.

If we are interested in the weighted average over *all* values of x, we must calculate the sum of the elements $x \cdot \Delta x \cdot f$ over the entire range of x values covered by the frequency distribution. Such a sum of infinitesimal elements is of course an integral. If A and B are, respectively, the lowest and the highest value of x, then this integral can be written in the form

$$\int_A^B x \cdot f(x)\, dx$$

where we have replaced Δx by the differential dx, and f by $f(x)$, since f is obviously a function of x. This integral is called the *expected value* of x, and denoted by the symbol $E(x)$. Thus, we have

$$E(x) = \int_A^B x \cdot f(x)\, dx \qquad (2.3)$$

It is readily apparent that the expected value is a weighted average of *all* values that x can take, each one weighted by its relative frequency. It is therefore nothing more than an extension of the simple concept of an arithmetic average to the *entire population* of values. We have defined it here for the case of a *continuous* variable. For discrete variables, Eq. (2.2) defines the expected value, provided that the summation covers *all* possible x values for which the frequency distribution is defined (i.e., the entire *population* of x values). The quantity $E(x)$ is also known as the *population mean*.

2.4 THE VARIANCE AS AN EXPECTED VALUE

If $g(x)$ represents any function of x (e.g., $g(x) = x^2 - 3x$), then the weight corresponding to any value of x is also the weight corresponding to the corresponding value of $g(x)$. Consequently, we define the expected value of $g(x)$ as

$$E\Big(g(x)\Big) = \int_A^B g(x) \cdot f(x)\, dx \qquad (2.4)$$

An expected value of special importance is that corresponding to the function

$$g(x) = [x - E(x)]^2 \tag{2.5}$$

The quantity $x - E(x)$ is of course analogous to the deviation $x - \bar{x}$, except that instead of $\bar{x}$, the sample mean, we now write $E(x)$, the population mean. The *population variance* is defined as the expected value of the function $g(x = [x - E(x)]^2$. Thus, writing Var(x) for population variance, we have

$$\mathrm{Var}(x) = E[x - E(x)]^2 = \int_A^B (x - E(x))^2 \cdot f(x)\, dx \tag{2.6}$$

where $f(x)$ is the frequency density of the variate x.

It is easy to see that Var(x) is an extension of the sample variance s^2, defined by Eq. (1.3), to the *entire population* of x-values. Indeed, $f(x)$ is the relative frequency for x in the population, analogous to $1/N$, the relative frequency of the value x in the sample. The distinction between N and $N - 1$ no longer matters, since for N very large (or infinite) the two values are essentially the same. Because Var(x) is the population analog to s^2, we call it the population variance and denote it usually by the symbol σ^2. Thus:

$$\sigma^2 = \int_A^B [x - E(x)]^2 \cdot f(x)\, dx \tag{2.7}$$

The square root of σ^2, namely σ, is called the *population standard deviation*. Let us illustrate the calculation of expected values for the rectangular distribution. We have seen (Fig. 2.6) that the frequency density function for the rectangular distribution is

$$f(x) = \frac{1}{\overline{AB}} = \frac{1}{B - A}$$

where A and B are the two extreme points of the range of x-values for which the distribution is defined. Equation (2.3) gives

$$E(x) = \int_A^B x \cdot f(x)\, dx = \frac{1}{B - A} \cdot \int_A^B x \cdot dx$$
$$= \frac{1}{B - A}\left[\frac{B^2}{2} - \frac{A^2}{2}\right] = \frac{A + B}{2}$$

Thus, as we certainly would have expected on intuitive grounds, the expected value of x is the midpoint of the segment AB.

Turning now to the variance, we obtain from Eq. (2.7):

$$\sigma^2 = \frac{1}{B - A}\int_A^B \left[x - \frac{A + B}{2}\right]^2 dx$$
$$= \frac{1}{B - A}\frac{1}{3}\left\{\left[B - \frac{A + B}{2}\right]^3 - \left[A - \frac{A + B}{2}\right]^3\right\}$$
$$= \frac{(B - A)^2}{12}$$

Therefore:

$$\sigma = \frac{B - A}{\sqrt{12}} \tag{2.8}$$

This is a useful result: for if we round a number like 8.633 to 8.63, we do it on the rationale that we are uncertain as to the correct value of the digit in the third decimal place. We believe, in other words, that the true value could have been anywhere between 8.625 and 8.635. Thus, the measured value would belong to a rectangularly distributed population over the range AB, where $A = 8.625$ and $B = 8.635$. We accept the value 8.630 as a plausible value and know, at the same time, that the uncertainty introduced by the rounding operation is measured by a standard deviation of $(8.635 - 8.625)/\sqrt{12}$, or 0.0029. These considerations are especially useful when our measured value is subject to other sources of uncertainty and we wish to compare the rounding error with the effects of these sources. A general rule is to round in such a fashion that the error caused by the rounding is negligible with respect to the errors due to these other sources. Failure to adhere to this rule may result in "rounding away" the variability in which one is interested. Examination of the scientific literature leads one to conclude that unfortunately such excessive rounding is more common than is usually believed. The result is that very useful information is then often irretrievably lost.

We briefly recapitulate:

1. We distinguish between a sample and a population of x-values, where x is a variate (also called a *random variable*).
2. A measure of the central value of a sample is the sample mean $\bar{x}$. For the population it is the population mean $E(x)$, often also denoted as μ.
3. A measure of width (or dispersion) for the sample is the sample variance s^2 given by Eq. (1.3). For the population it is given by the population variance $Var(x)$ or σ^2, given by Eqs. (2.6) or (2.7). Since s^2 and σ^2 do not have the dimension of x (but rather that of x^2), we also consider the *standard deviations* s and σ, which are the square roots of s^2 and σ^2, respectively. The quantities s and σ have the dimension of x and are therefore expressed in the same units as x.
4. The *population parameters* μ and σ are fixed numbers (generally unknown). This is not the case of $\bar{x}$ and s, which vary from sample to sample, even when the samples are all taken from the same population. This is expressed by the statement that x and s are subject to *sampling fluctuations* (also known as sampling errors, an expression we do not recommend).
5. Since μ and σ are generally unknown, all we have in practice are $\bar{x}$ and s, which can be considered as *estimates* (or better, *sample estimates*) of μ and σ, respectively. As the sample size N increases, the estimates $\bar{x}$ and s increasingly tend to approach the corresponding population parameters μ and σ.

2.5 CHANGES OF SCALE, CODING, STANDARD VARIATE

A pertinent question is what effect a change of units or scale has on the parameters of a distribution and on their estimates.

It is easy to see that if x is multiplied by a constant factor k (for example, $k = 2.54$ to change from inches to centimeters), μ, σ, $\bar{x}$, and s will all be multiplied by k, too. Consequently, σ^2 and s^2 will be multiplied by k^2.

Adding (or subtracting) a constant B, on the other hand will affect the means μ and $\bar{x}$, but leave the standard deviations unchanged.

For example, changing temperature units from Celsius to Fahrenheit,

$$F = 32 + 1.8C$$

will affect the means as follows:

$$\mu_F = 32 + 1.8\mu_C \qquad \bar{x}_F = 32 + 1.8\,\bar{x}_C$$

and the standard deviations according to

$$\sigma_F = 1.8\sigma_C \qquad s_F = 1.8s_C$$

It is often advantageous to change the scale *temporarily* in order to simplify calculations. Suppose, as an example, that we are required to calculate the mean and the standard deviation of the sample

$$x = (1082.64, 1083.18, 1082.85, 1083.07)$$

Let us consider a variable u defined by

$$u = 100 \cdot (x - 1082)$$

so that

$$x = 1082 + 0.01 \cdot u$$

Then we obtain

$$u = (64, 118, 85, 107)$$

We find readily, for the mean and standard deviation of u:

$$\bar{u} = 93.5 \qquad s_u = 23.98$$

Then, using the rules given above, we have

$$\bar{x} = 1082 + 0.01(93.5) = 1082.935$$

and

$$s_x = 0.01s_u = 0.2398$$

What we have just done is referred to as *coding* the x-values for the purpose of simplifying the calculations. Even when using a calculator or a computer, coding helps in eliminating or greatly reducing rounding errors.

A special change of scale, *not* motivated by a change of units (such as Celsius to Fahrenheit) is of great importance in statistics. This is *reduction of a variate to standardized form.*

Taking again the sample of four x-values above, let us make the transformation

$$z = \frac{x - \bar{x}}{s_x} \tag{2.9}$$

This is of the form discussed above since we can write it as

$$z = \frac{1}{s_x}(x - \bar{x})$$

Our rules tell us that

$$\bar{z} = (\bar{x} - \bar{x})\frac{1}{s_x} = 0$$

$$s_z = \frac{s_x}{s_x} = 1$$

Thus, the standardized variable z has *zero-mean* and *unit-standard deviation*. It is easily seen that had we operated on the coded values u above, we would have obtained the same z-values. In either case, we have

$$z = (-1.2302,\ 1.0217,\ -0.3545,\ 0.5630)$$

and we can readily verify that $z = 0$ and $s_z = 1$. If the population mean and the population standard deviation are known, reduction to standard form can be defined as

$$z = \frac{x - \mu}{\sigma} \tag{2.10}$$

Then the *population mean* of z will be zero and its *population standard deviation* will be unity. Sometimes this transformation is useful as a theoretical construct, even when μ and σ are not known.

2.6 COVARIANCE, CORRELATION, STATISTICAL INDEPENDENCE

In mathematics, two variables x and y may be related through a functional equation such as $y = 2 - 3x$, or $y = x^2$, or $y = \log(3 + x)$. In such cases, a plot of y versus x will be a straight line or a smooth curve. For example, if x represents the radius of a circle and y its area, then $y = \pi \cdot x^2$. In order to observe the relationship, it is then necessary to consider a number of circles of different radii.

In statistics we deal with a different situation. The relationships we consider here are between *fluctuations* rather than between the variables themselves. Taking the same example, let us assume that every time we measure the radius of a circle, our measurement is affected by a measurement error, say δ, so that if R is the true value of radius, the measurement x will be

$$x = R + \delta \tag{2.11}$$

Similarly, the *measurement* y of the area (*not* the calculated value πx^2) will be affected by a measurement error, say ε, so that for a true value of the area equal to A, we measure

$$y = A + \varepsilon \tag{2.12}$$

If we consider a collection of circles of different radii, then obviously y will be related to x and we will observe this functional relation in a plot of y versus x. But our present interest is not in this relation but rather in that existing between δ and ε. It now does not matter very much whether our series of measurements is made on a single circle or on a set of circles of different radii. We visualize a sample of *replicate* measurements $x_1, x_2, \ldots, x_N$ and a sample of associated measurements $y_1, y_2, \ldots, y_N$. Let us assume that the object we are measuring is a single circular disk of known thickness t and specific gravity d. Then the radii x could be measured by an optical device, and the areas y by weighing the disk, and calculating the area from the weight W by means of the formula $y = W/(t \cdot d)$. Because of random

Table 2.1 Calculation of Covariance

Lab	x	y	$x - \bar{x}$	$y - \bar{y}$	$(x - \bar{x}) \cdot (y - \bar{y})$
1	10.757	16.750	0.356	0.389	0.13848
2	9.980	16.083	−0.421	−0.278	0.11704
3	10.117	16.010	−0.284	−0.351	0.09968
4	10.723	16.770	0.322	0.409	0.13170
5	10.377	15.607	−0.024	−0.754	0.01810
6	9.587	14.940	−0.814	−1.421	1.15669
7	11.267	18.367	0.866	2.006	1.73720

$\bar{x} = 10.4011$ $\bar{y} = 16.3610$ $\frac{\Sigma(x - \bar{x})(y - \bar{y})}{N - 1} = 0.56648$

fluctuations in both measuring processes, neither the x nor the y will be the same from measurement to measurement. Writing the two series of measurements

$$x_1, x_2, \ldots, x_N$$

$$y_1, y_2, \ldots, y_N$$

we will, in all likelihood, observe *no* meaningful relation between the x and the y. The reason is that the measurement processes are totally unrelated: the *fluctuation* ε_i in the measurement y_i is totally unrelated to the *fluctuation* δ_i in the measurement x_i. We will say that x and y are *statistically independent*.

Let us now consider a different case. In Table 2.1, x and y are measurements of pentosans made by seven different laboratories on two pulps, which we will call P1 and P2.* Ideally, all seven values for P1 should be the same, since they were obtained using the same material. The same holds for the seven values for P2. However, because of experimental error, the values differ somewhat from each other, both for P1 and for P2. The question we ask is whether the

*x and y are the averages of the triplicate measurements in columns H and I of Table 1.2.

fluctuations in P1 and P2 are related to each other. Suppose that because of slight differences in the measuring set-up in the seven laboratories, one laboratory may tend to get low values, while another laboratory tends to get high values. The causes for such an occurrence are called "systematic errors" in the laboratories. Then to a low value for P1 there should correspond a low value for P2, and similarly for high values. To test whether this is the case, we have calculated, for each of the 14 values, the corresponding deviations from the averages, $x - \bar{x}$ and $y - \bar{y}$. A glance at these deviations convinces us that if a laboratory tends to be low (high) for P1, it will also tend to get low (high) values for P2. In column 6 of Table 2.1, we have listed the product of the deviation of x by the corresponding deviation of y. These products have been summed and divided by $N - 1$. The result, given at the bottom of the column is called the *covariance* of the sample of paired x and y values. If the association between x and y is strong, the covariance will be large. If the association is weak or nonexistent, the covariance will be closer to zero.

The covariance depends not only on the degree of association between x and y, but also on their respective magnitudes or, more precisely, on the magnitude of the deviations $x - \bar{x}$ and $y - \bar{y}$. Therefore, a standardized form of the covariance has been devised, called the *correlation coefficient* between x and y. It is obtained by summing the products, not of $x - \bar{x}$ with $y - \bar{y}$, but rather of z_x by z_y, where the z are the x or y expressed in standard form (see Eq. (2.10)). The calculations are shown in Table 2.2 for the x, y of Table 2.1.

The mathematical expressions for the covariance and the correlation coefficient, denoted by the symbols cov(x,y) and $r(x,y)$, are

$$\mathrm{cov}(x,y) = \frac{\sum [x_i - \bar{x}][y_i - \bar{y}]}{N - 1} \tag{2.13}$$

$$r(x,y) = \frac{1}{N - 1} \sum \left[\frac{x_i - \bar{x}}{s_x} \cdot \frac{y_i - \bar{y}}{s_y} \right] \tag{2.14}$$

$$r(x,y) = \frac{1}{N - 1} \frac{\sum [x_i - \bar{x}][y_i - \bar{y}]}{s_x \cdot s_y} \tag{2.15}$$

Table 2.2 Calculation of Correlation Coefficient (Data of Table 2.1)

Lab	x	y	z_x	z_y	$z_x \cdot z_y$
1	10.757	16.750	0.6323	0.3568	0.2256
2	9.980	16.083	−0.7481	−0.2550	0.1908
3	10.117	16.010	−0.5047	−0.3220	0.1625
4	10.723	16.770	0.5719	0.3752	0.2145
5	10.377	15.607	−0.0428	−0.6916	0.0296
6	9.587	14.940	−1.4463	−1.3034	1.8851
7	11.267	18.367	1.5383	1.8400	2.8305

$\bar{x} = 10.4011 \quad \bar{y} = 16.3610 \qquad r(x,y) = \dfrac{\Sigma\, z_x \cdot z_y}{N - 1} = 0.9231$

$s_x = 0.5629 \quad s_y = 1.0902$

It is important to observe that both the covariance and the correlation coefficient, as defined by Eqs. (2.13) and (2.14), are *sample estimates*. A new sample of paired x and y values, for example, from a set of seven different laboratories, would yield different values for cov(x,y) and $r(x,y)$. Here again we conceptually extend these two new quantities to what we would obtain by taking a larger and larger sample of paired x and y values. In the limit, for the sample size N tending to infinity, the covariance as well as the correlation coefficient would approach fixed values, which we denote by Cov(x,y) and $\rho(x,y)$, and which we call the *population covariance* and the *population correlation coefficient*. It is not difficult to understand that they can be expressed by using the concept of the expected value, as follows:

$$\text{Cov}(x,y) = E[(x - E(x) \cdot (y - E(y)] \tag{2.16}$$

$$\rho(x,y) = \frac{E[(x - E(x) \cdot (y - E(y)]}{\sqrt{E[x - E(x)]^2 \cdot E[y - E(y)]^2}} \tag{2.17}$$

In contradistinction to cov(x,y) and r(x,y), the quantities Cov(x,y) and $\rho(x,y)$ are fixed numbers, not depending on the particular sample of x,y pairs of the experiment. They are, in other words, *population parameters*, of which cov(x,y) and $r(x,y)$ are *sample estimates*. Note that

$$r_{x,y} = \frac{\text{cov}(x,y)}{s_x \cdot s_y} \tag{2.18}$$

$$\rho(x,y) = \frac{\text{Cov}(x,y)}{\sigma_x \cdot \sigma_y} \tag{2.19}$$

Returning to Table 2.2, we find $r(x,y) = 0.9230$. This is a value close to $+1$ and indicates a fairly strong "positive correlation" between x and y. There are cases in which x and y tend to vary in opposite direction: if x increases, y decreases. For such situations, the correlation coefficient will be negative, and if the (inverse) association is strong, the correlation coefficient will be close to (-1). It can be proved that both r and ρ always lie between the values (-1) and $(+1)$. A value close to zero generally indicates a weak association, but this statement must be qualified, as we will shortly see. It can also be shown that if $\mathrm{r} = 1$, then a plot of y versus x will yield points falling exactly on a straight line of positive slope. For $r = 1$, the points also fall exactly on a straight line, but the slope will be negative.

It is important to note that the correlation coefficient is a measure of *linear* association, and only of linear association. To illustrate this point, consider the relation

$$x^2 + y^2 = R^2$$

which is the equation of a circle of radius R whose center is the origin of the coordinate system. If we plot the *upper half* of this circle, from the point $x = -R$, $y = 0$, to the point $x = +R$, $y = 0$, there is of course a smooth functional relation between the x and the y. Nevertheless, the correlation coefficient for the totality of points on this half-circle is zero. We will see why this is so when we study the fitting of straight lines. Suffice it to say here that the relation, in spite of the zero-correlation, is not at all weak. It is in fact quadratic, not linear; in this case, the zero correlation indicates nonlinearity, not lack of association.

Apart from the fact that the correlation coefficient is a measure of *linear* association only, it should also be noted that the practice of studying relationships between measurements of different properties by means of correlation coefficients is to be condemned. For example, if we wish to study the relationship between the volume V and the pressure P of a given mass of gas at constant temperature, and our data consisted of pairs of values V, P, the proper procedure is to start by plotting P versus V on graph paper or on a computer display, and observing the relation obtained. It would be *totally inappropriate* to attempt to measure the "strength of association" between volume and pressure by calculating the correlation coefficient between V and P. As we have seen, the correlation coefficient is useful as a measure of the association between *random fluctuations*. Physical or chemical relationships should be studied by more appropriate techniques than by the display of correlation coefficients.

Finally, we mention the following fact. If two random variables x and y are *statistically independent,* then the correlation coefficient $\rho(x,y)$ is zero. The converse is not always true, but we will see that under certain conditions it *is* true.

2.7 THE COMBINATION OF RANDOM VARIABLES

We have already observed that most properties of interest in science are measured *indirectly*: they often are *functions* of one or more direct measurements. In gravimetric analysis, for example, a sample is weighed out. After chemical treatment, a compound is derived from it and weighed. If the weight of the sample is S and the weight of the compound C, then the "unknown" is calculated as

$$x = K \frac{C}{S} \tag{2.20}$$

where K is a known numerical constant. The direct measurements are S and C. The unknown x is a *derived* measurement. The problem is to determine the uncertainty of the calculated (derived) quantity x, given the uncertainties of C and S.

Let us start with a simpler case. Suppose that a chemical compound has four constituents (for example, iron, carbon, cobalt, and nickel) and that all but the carbon can be readily measured. Now we have for the percentage contents in the sample:

$$\%\text{Fe} + \%\text{C} + \%\text{Co} + \%\text{Ni} = 100$$

and consequently we can estimate the carbon content from:

$$\%\text{C} = 100 - \%\text{Fe} - \%\text{Co} - \%\text{Ni}$$

or, representing %C by x, %Fe by y, %Co by z, and %Ni by u:

$$x = 100 - y - z - u \tag{2.21}$$

The problem is to determine the uncertainty of x (the *calculated quantity*) from that of the measurements y, z, and u.

2.7.1 Linear Combinations

Equation (2.21) is *linear* in x, y, z, and u. More generally, we consider linear combinations given by the relation

$$x = b_0 + b_1 \cdot y_1 + b_2 \cdot y_2 + b_3 y_3 + \cdots \tag{2.22}$$

where $y_1, y_2, \ldots,$ are the measurements; $b_1, b_2, \ldots,$ are known numerical constants, and x is the quantity of interest. It can be shown that the variance of x in Eq. (2.22) is given by the general formula:

$$\begin{aligned} \text{Var}(x) = \; & b_1^2 \,\text{Var}(y_1) + b_2^2 \,\text{Var}(y_2) + b_3^2 \,\text{Var}(y_3) + \cdots \\ & + 2\, b_1 b_1 \,\text{Cov}(y_1, y_2) + 2\, b_1 b_3 \,\text{Cov}(y_1, y_3) \\ & + 2\, b_2 b_3 \,\text{Cov}(y_2, y_3) + \cdots \end{aligned} \tag{2.23}$$

An important special case is that in which the fluctuations in the measurements $y_1, y_2, y_3, \ldots$ are all statistically independent of each other. In that case the covariances are all zero and we have

$$\begin{aligned} \text{Var}(x) = \; & b_1^2 \,\text{Var}(y_1) + b_1^2 \,\text{Var}(y_2) \\ & + b_3^2 \,\text{Var}(y_3) + \cdots \end{aligned} \tag{2.24}$$

Note that in combining measurments, it is not the standard deviations that are additive, but rather the squares, that is, the variances.

A simple illustration of Eq. (2.23) is given by the *difference* of two independent measurements:

$$x = y_1 - y_2$$

This can be written as

$$x = 1 \cdot y_1 + (-1) \cdot y_2$$

Hence:

$$\begin{aligned} \mathrm{Var}(x) &= (1)^2 \,\mathrm{Var}(y_1) + (-1^2 \,\mathrm{Var}(y_2) \\ &= \mathrm{Var}(y_1) + \mathrm{Var}(y_2) \end{aligned}$$

Note that the variance of the difference is *not* the difference of the variances, but rather their sum.

2.7.2. Multiplicative Combinations

Equation (2.20) is not linear and therefore not covered by Eqs. (2.23) and (2.24). It is however multiplicative. A more general case is given by the relation

$$x = Ky_1^{a_1} \cdot y_2^{a_2} \cdot y_3^{a_3} \tag{2.25}$$

where a_1, a_2, a_3 are exponents of known numerical value. If y_1, y_2, y_3, . . . are all mutually statistically independent, then the general formula is

$$\frac{\mathrm{Var}(x)}{x^2} = a_1^2 \frac{\mathrm{Var}(y_1)}{y_1^2} + a_2^2 \frac{\mathrm{Var}(y_2)}{y_2^2} + a_3^2 \frac{\mathrm{Var}(y_3)}{y_3^2} \tag{2.26}$$

This relation can be written in the form

$$\frac{\sigma_x}{x} = \sqrt{a_1^2 \left[\frac{\sigma y_1}{y_1}\right]^2 + a_2^2 \left[\frac{\sigma y_2}{y_2}\right]^2 + a_3^2 \left[\frac{\sigma y_3}{y_3}\right]^2} \tag{2.27}$$

The ratio of a standard deviation of a variable to the value of this variable is called the *coefficient of variation* of this variable.* Chemists

*Many authors define the coefficient of variation as 100 σ_x/x. We prefer to call this quantity the "percent coefficient of variation."

prefer to call it the *relative standard deviation*. Thus, for equations of the multiplicative type of Eq. (2.25), it is not the squares of the standard deviations that are additive, but rather the squares of the *relative standard deviations*.

Returning, as an illustration, to Eq. (2.19), we obtain

$$\left[\frac{\sigma_x}{x}\right]^2 = \left[\frac{\sigma_c}{c}\right]^2 + \left[\frac{\sigma_s}{s}\right]^2$$

2.7.3 Logarithmic and Exponential Functions

Finally, we give formulas for functions involving logarithms and exponentials.

(a) For the *natural* (base e) logarithm:

$$x = \ln y \tag{2.28}$$

the formula is

$$\sigma_x = \frac{\sigma_y}{y} \tag{2.29}$$

(b) For

$$x = e^y \tag{2.30}$$

we have the formula

$$\sigma_x = e^y \cdot \sigma_y \tag{2.31}$$

A general formula, covering all the preceding cases, as well as functions of any other type, can be derived; the derivation is based on the following properties of expected values:

$$E(x + y + z + \cdots) = E(x) + E(y) + E(y) + E(z) + \cdots \tag{2.32}$$

$$E(k \cdot x) = k \cdot E(x) \tag{2.33}$$

$$E(k) = k \tag{2.34}$$

where k is a constant. Equation (2.32) is always valid, even when the variates $x, y, z, \ldots$ are not statistically independent.

2.8 SUMMARY

In this chapter we have discussed the basic concept of a frequency distribution and introduced the important concept of an expected value. We have learned to differentiate between population parameters and their sample-estimates. Exact definitions have been given for mean, standard deviation, and variance, both for a sample and for a population. The correlation coefficient has been defined, and we have explained the law of propagation of errors, i.e., the law by which the variability of derived measurements is calculated.

3

Precision and Accuracy: The Central Limit Theorem, Weighting

3.1 SCOPE

The concepts of precision and accuracy are of great practical importance in the evaluation of measurements. Of fundamental importance is the central limit theorem that justifies the weight attached to the normal distribution statistics. This chapter also deals with the subject of weighting in the analysis of experimental data.

3.2 THE STANDARD ERROR OF THE MEAN: PRECISION AND ACCURACY

Referring to Fig. 1.3, we recall that the frequency diagram to the right is generated by the succession of measurements x_i. Suppose that we group the 25 measurements into groups of five consecutive values and that we calculate the average $\bar{x}$ of each group of 5. This yields 5 averages $\bar{x}_1$, $\bar{x}_2$, $\bar{x}_3$, $\bar{x}_4$, and $\bar{x}_5$. We now visualize again an unlimited continuation of this process, by postulating first our unlimited sequence of individual measurements, subsequently grouped as we have indicated, thus yielding an unlimited sequence of $\bar{x}$-values (each $\bar{x}$ being an average of 5 consecutive x-values). The $\bar{x}$-values will then generate a frequency distribution of their own. We will discuss in the next section the nature of this "frequency distribution of averages." Here we concern ourselves with only two questions: "What is the expected value of this new distribution?" and "What is the variance of this new distribution?" We assume for the moment that we know the mean μ and variance σ^2 of the original distribution "of single measurements."

Using Eqs. (2.32) and (2.33), we have

$$E(x) = E\left(\frac{x_1 + x_2 + x_3 + x_4 + x_5}{5}\right)$$
$$= \frac{1}{5}[E(x_1) + \cdots + E(x_5)]$$

But $x_1, x_2, \ldots, x_5$ are all members of the same population, of expected value $E(x)$. Hence:

$$E(x) = \frac{1}{5} \cdot 5\ E(x) = E(x)$$

In general we have

$$E(\bar{x}) = E(x) \tag{3.1}$$

as we might well have guessed on intuitive grounds.

Furthermore, using Eq. (2.23), which is legitimate because x_1, $x_2, \ldots$ are all independent drawings from the same population, we have

$$\text{Var}(\bar{x}) = \text{Var}\left(\frac{x_1 + \cdots + x_5}{5}\right)$$
$$= \frac{1}{5^2}\,\text{Var}(x_1) + \cdots + \frac{1}{5^2}\,\text{Var}(x_5)$$

hence,

$$\text{Var}(\bar{x}) = \frac{1}{25} \cdot 5 \cdot \text{Var}(x) = \frac{\text{Var}(x)}{5}$$

More generally, for the average $\bar{x}$ of N independent measurements from the same population, we have

$$\text{Var}(\bar{x}) = \frac{\text{Var}(x)}{N} \tag{3.2}$$

and consequently

$$\sigma_{\bar{x}} = \frac{\sigma_x}{\sqrt{N}} \tag{3.3}$$

The standard deviation $\sigma_{\bar{x}}$ is often called the *standard error of the mean,** and Eq. (3.3) expresses the law of the standard error of the mean.

Equation (3.3) is an important relation with important practical consequences. It tells us that the variability of averages of several replicate measurements is *smaller* than that of the original measurements. It confirms our intuitive feeling that an average of several measurements is "better" than a single measurement. At the same time it tells us that the improvement, in terms of *reduction of variability,* is slow; for example, it takes 16 measurements to reduce the standard deviation by four.

A measurement process that has *small variability* (as expressed, for example, by a small standard deviation) is said to have *high precision.* Of course, "small" and "high" are relative concepts, requiring a standard of comparison to become meaningful. But at any rate, *precision* is an expression of relative smallness of variability, or in other words, of relative stability within the measuring process. A measuring process of high precision is not necessarily a satisfactory one. An additional desideratum is that the measurements yielded by

*The expression "standard error" is often used for the standard deviation of a *derived* quantity, such as $\bar{x}$. There is however no objection to calling it simply the standard deviation of the mean. The distinction is gradually disappearing from the statistical vocabulary.

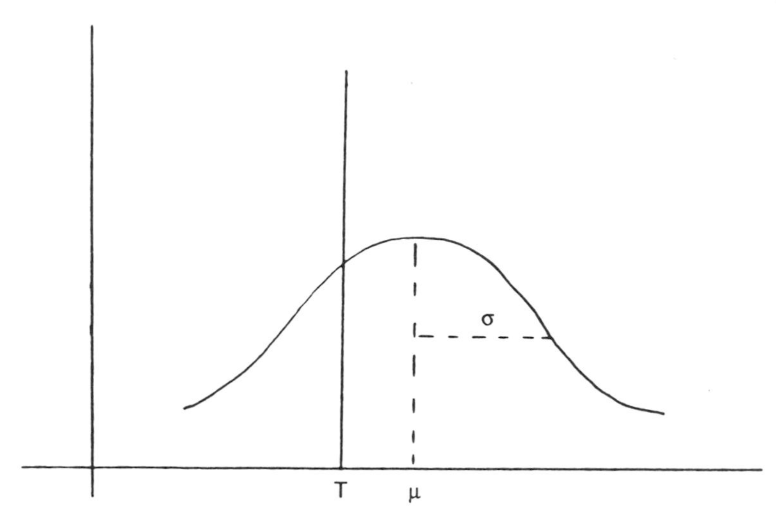

Figure 3.1 Precision and accuracy.

the process also be close to the "true value" of the measured property. The matter is illustrated in Fig. 3.1. The mean of the population is μ and its standard deviation is σ. If μ is also the "true value" of the property measured, then "on the average," the process will yield the true value. We call such a process "unbiased." On the other hand, it may happen that the true value is distinctly different from the population mean μ, and equal, say, to T. Then $\mu - T$ is the "bias" of the process for the measurement of the property in question.

Accuracy is a generic concept expressing smallness of "total error." Suppose that the value for some single measurement is M. Since the true value is T, the "total error" is $M - T$. But $M - T$ can be partitioned into two parts:

$$M - T = (M - \mu) + (\mu - T) \tag{3.4}$$

The second, $\mu - T$ is simply the bias of the process. It cannot be reduced, or eliminated, without changing the process itself. On the other hand, the first part, $M - \mu$ is the "random" deviation of the measurement from its mean μ. If we repeated the measurement M, say nine times, and took the average M of the nine measurements, the mean of M would still be μ, and the bias consequently unchanged. But the random part, $M - \mu$, being a member of a population of deviations of standard deviation $\sigma/\sqrt{9}$, or $\sigma/3$, would, on the average, be three times smaller than for a single measurement M. In the limit, for an average M of a very large number N of replicate measurements, the random part $M - \mu$ would eventually tend to zero (because the standard error $\sigma/\sqrt{N}$ would tend to zero), and the remaining error would be the pure bias $\mu - T$. Some authors then measure the *accuracy of the process itself* (in contradistinction to the accuracy of a single measurement or of the average of a small number of measurements) by the bias $\mu - T$: the smaller this bias, the more accurate will be the method.

While the concept of precision is of supreme importance for the evaluation of a measuring process, the concept of accuracy has been gradually diminishing in importance in recent times. There are two major reasons for this. In the first place, as we have seen, few measurements are "direct" measurements of a property of interest. Generally, the measurement is a "related" quantity (for example, the intensity of a spectrum at a specified wavelength), obtained by an instrumental method of an often complicated nature. Secondly, and as a result of the first reason, a measurement can never be accepted as valid, unless the equipment used in obtaining it has been *calibrated* recently enough to ensure that the calibration is still valid. If the measuring process has been properly calibrated, and one is confident that the most recent calibration is still valid, then the concept of *accuracy* (in terms of *bias*) is of limited relevancy. Because of the calibration process, the bias will be zero or, at any rate, quite small. Thus, the concern for accuracy has now shifted to one for proper calibration, a subject that will be discussed in a subsequent chapter.

3.3 THE CENTRAL LIMIT THEOREM

This book does not, in general, present theorems. Indeed, it is not our intent to write a textbook on mathematical statistics, but rather to develop only those concepts and mathematical relations that are needed for carrying out a meaningful analysis of experimental data. We now make an exception, because the theorem we wish to discuss is of great practical significance.

In the preceding section we have shown that the mean of a distribution of averages $\bar{x}$ is the same as that of the original measurements x. We have also shown that the standard deviation of $\bar{x}$ is *smaller* than that of x, by a factor of $\sqrt{N}$, where N is the number of original measurements of which $\bar{x}$ is the average. It is worth emphasizing that these two facts were proved without even mentioning the type of distribution considered: rectangular, triangular, etc. The reason is that both facts are valid for *any* distribution. The question we now consider is whether the process of averaging has any effect on the type of distribution. More specifically, if we start with a population of measurements x of a particular type of distribution, will the population of $\bar{x}$ be of the same type, or will it be different? And if it is different, can we say anything specific about it?

The answer to these questions is a remarkable statistical fact, called the *central limit theorem*. Stated without complete rigor, it says that no matter what type of distribution we start with, the distribution of averages is always of one specific type, called the *normal* or *Gaussian* distribution. Theoretically, this statement is only true *in the limit,* when $\bar{x}$ is the average of a very large number of original measurements. However, while this qualification is required theoretically, it turns out that the "approach to normality" implied by the theorem is in practice very rapid, so that even for $N = 4$, $\bar{x}$ is essentially normal, regardless of what the distribution of x is.

In the next section, we discuss the nature of the normal distribution. But first we issue an important warning. The averaging process with which the theorem deals has to be of a *random* nature. We cannot "pick" our x values before averaging, but must carry out the averaging on *random samples* from the original distribution. Only then will the theorem apply. We will illustrate the theorem after discussing the normal distribution.

3.4 THE NORMAL DISTRIBUTION

Every continuous distribution is characterized by its frequency density. The frequency density of the normal distribution is given by the formula

$$f(x) = \frac{1}{\sigma\sqrt{2\pi}} \exp\left[-\frac{(x-\mu)^2}{2\sigma^2}\right] \tag{3.5}$$

Note that $f(x)$ contains two parameters, μ and σ^2. It can be shown that the mean of the distribution of which Eq. (3.5) is the frequency density is indeed μ and that its variance is indeed σ^2, provided that the distribution extends over a range of x-values from $(-\infty)$ to $(+\infty)$. Thus we have, for the normal distribution:

$$E(x) = \frac{1}{\sigma\sqrt{2\pi}} \int_{-\infty}^{+\infty} x \cdot \exp\left[-\frac{(x-\mu)^2}{2\sigma^2}\right] dx = \mu \tag{3.6}$$

$$\text{Var}(x) = \frac{1}{2\pi\sigma^2} \int_{-\infty}^{+\infty} (x-\mu)^2 \exp\left[-\frac{(x-\mu)^2}{2\sigma^2}\right] dx = \sigma^2 \tag{3.7}$$

A graph of the normal distribution is shown in Fig. 3.1. Noteworthy is the fact that there is complete *symmetry* with respect to μ: the figure on the left of μ is the mirror image of the figure on its right. This is also apparent from Eq. (3.5), because the density is the same for the two symmetrical points $x_1 = \mu + d$ and $x_2 = \mu - d$, for any d.

It can also be shown that the points on the curve for which the abscissa values are $\mu - \sigma$ and $\mu + \sigma$, are the *points of inflexion,* that is, the points at which the curvature of the curve changes from convex to concave (with respect to the x-axis), or vice versa. It is also seen that if one considers points for which the abscissa is $\mu - 3\sigma$ and $\mu + 3\sigma$, these points will be far "in the tail" of the distribution, that is, in a region for which the frequency density is very small. It can be shown that about 95% of the distribution will be concentrated in the region extending from $\mu - 2\sigma$ to $\mu + 2\sigma$.

The normal distribution has been tabulated in a number of ways. Before discussing this subject, we consider the "standard" normal distribution. In accordance with Eq. (2.9), it is given by:

$$z = \frac{x - \mu}{\sigma} \tag{3.8}$$

As we have seen, we have for the distribution in standard form:

$$E(z) = 0 \tag{3.9}$$

$$\mathrm{Var}(z) = 1 \tag{3.10}$$

Normal distributions vary from each other only in terms of μ and/or σ. They all have the same standard form, whose frequency density is

$$f(z) = \frac{1}{\sqrt{2\pi}} e^{-(z^2/2)} \tag{3.11}$$

Consequently, it is necessary to tabulate only the standard normal distribution. We will show how this table can be used in connection with any normal distribution.

The importance of the normal distribution derives to a large extent from the central limit theorem, since the use of averages is very common in all scientific work.

Tables of the standard normal distribution, generally denoted by z, come in a variety of forms. Because of the small density in the tails, it is seldom necessary to consider values of z less than -4, or greater than $+4$. Consequently, the range in the tables is generally -4 to $+4$. Many tabulations give the "ordinate" at each tabulated z-value. This is simply the density given by Eq. (3.11). It is readily calculated, even on a handheld calculator, for any value of z. Therefore, tabulation of the ordinate is essentially unnecessary. On the other hand, the "area" is not easily calculated and therefore requires tabulation. In this book, a table appears in the Appendix titled "Areas of the Standard Normal Distribution." For each positive z-value, from $z = 0$ to $z = +4$, we give the area under the curve between the vertical line going through $z = 0$ and the vertical line going through the value of z in question. This is the shaded area shown at the beginning of the table. It is the "probability" (or relative frequency) of obtaining at random a value between 0 and the considered value of z.

We see that for $z = 1$, the tabulated area is 0.3413. Hence $2 \times .3413$ or about 2/3 of all values lie between $z = -1$ and $z = +1$.

For $z = 2$, the tabulated value is .4772. Hence about 95% of all values lie between $z = -2$ and $z = +2$.

For $z = 3$, the tabulated value is .4987. Hence about 99.7% of all values lie between $z = -3$ and $z = +3$.

These results can now be used (as can the entire table) for *any* normal distribution. Consider a normal population of mean μ and standard deviation σ. Then the relative frequency corresponding to an interval

$$A \leqslant x \leqslant B$$

is exactly the same as the relative frequency corresponding to

$$\frac{A - \mu}{\sigma} \leqslant \frac{x - \mu}{\sigma} \leqslant \frac{B - \mu}{\sigma}$$

or

$$\frac{A - \mu}{\sigma} \leqslant z \leqslant \frac{B - \mu}{\sigma}$$

Knowing A, B, μ, and σ, this latter frequency is readily obtained from the table. It is then also the desired frequency corresponding to the interval $A \leqslant x \leqslant B$.

As an example, consider a normal population x of mean $\mu = 5$ and standard deviation $\sigma = 1.5$. What is the relative frequency corresponding to the interval $0 \leqslant x \leqslant 8.2$?

We have

$$\frac{0 - 5}{1.5} \leqslant z \leqslant \frac{8.2 - 5}{1.5}$$

or

$$-3.33 \leqslant z \leqslant 2.13$$

The interval (-3.33 to 2.13) is the sum of the intervals (-3.33 to 0) and (0 to 2.13). Let F1 and F2 represent the frequencies corresponding to these intervals. For F2 we find at once from the table: F2 = .4834. As to F1, this is of course identical with the frequency for the interval (0 to 3.33), for which we find from the table F1 = .4996. Hence the desired frequency is F1 + F2 = .4996 + .4834 = .9830.

It is useful to memorize the following three items. In a normal population: (a) A ± "1-sigma" interval around the mean contains about two-thirds of the population; (b) A ± "2-sigma" interval around the mean contains about 95% of the populations; and (c) A ± "3-sigma" interval around the mean contains about 99.7% of the population, which is very close to the entire population.

3.5 POOLING SAMPLE VARIANCES

We have seen that a sample estimate s^2 of a variance, is a random variable, subject to fluctuations from sample to sample. The magnitude of these fluctuations is governed, other things being equal, by the degrees of freedom on which the estimate is based. The smaller the number of degrees of freedom, the larger will be the fluctuations of a series of estimates s^2 resulting from independent samples taken from the same population. Consequently, if only a single such estimate is available, its "uncertainty" will be larger to the extent that the degrees of freedom are smaller. This poses a real dilemma: in order to attain a reasonable degree of "stability" (i.e., of smallness of uncertainty) in an estimate s^2, it should be based on about 30 degrees of freedom or more. In order to achieve this, about 30 replicate measurements have to be made. In most practical work, this is a prohibitively large number of replicates. Chemical laboratories will make two or three determinations in routine analyses, and not many more in research situations. Some instrumental determinations can be made more expeditiously and hence tolerate economically, a larger number of replicates. However, in many of these situations, the instrumental measurement is only the last step of a sequence of operations, and such that the preceding steps are far more elaborate, requiring "wet chemistry." A repetition of the entire analysis is therefore again subject to severe economic and time limitations. The dilemma can often be solved through the following considerations. While a particular chemical sample submitted for analysis to a commercial laboratory is seldom given more than duplicate analyses, it is also true that many such samples will be submitted in the course of a week, a month, or a year. This suggests that means should be found to *combine* the evidence from many such samples, to calculate a more stable estimate of variability.

Let us therefore start by trying to enumerate the factors that will influence the variance estimate s^2 obtained from N measurements x_1, x_2, . . . , x_N.

One of these factors is undoubtedly the number of degrees of freedom, $N - 1$. Another factor results from the *actual* differences in the property measurement between the N individual test specimens submitted to the analysis. If the material is very homogeneous, the variability between the N results will be the result of measurement errors only. If not, this variability will include in addition to measurement error the actual differences between the test specimens. This cannot be avoided in general and in such cases, the experimenter realizes that s^2 measures *not* pure measurement error, but a combination of "heterogeneity" of the material and measurement error.

It is also a fact that the magnitude of x is often a factor affecting the magnitude of s^2. For example, a sample of animal feed that contains 60% protein may show more variability in replicate measurements of its protein content than a sample that contains only 15% protein.

Finally, even when we consider a set of samples from the same *type* of material, say animal feeds, some of these may have a different composition than others. Thus, the fact that two such materials may have approximately the same protein content, does not necessarily ensure their similarity with respect to the variability encountered in analyzing them for protein. Chemists express this by talking about "interfering" substances or constituents, or by stating that the (chemical) *matrix* of the two materials is different.

Returning now to the idea of combining the evidence from many *sets* of replicate measurements obtained on different but similar materials, we can formulate a possible approach as follows.

In the notebook of a particular routine analytical (or clinical) laboratory, we may find a collection of results, of duplicate or triplicate (or even more) measurements accumulated in the course of time. If all of these sets were obtained on materials with the same, or approximately the same, matrix, and all had the same, or approximately the same, value of the property measured, then we should be able to combine their evidence to obtain a better estimate of variability. Table 3.1 is an illustration of such a situation. It consists of seven sets of measurements for the determination of milkfat in samples of packaged milk. For each set, we tabulate, in addition to the original measurements, the mean, the sum of squares SS, the degrees of freedom DF, and the sample variance s^2. We have in accordance, with Eq. (1.7), $s^2 = \text{SS/DF}$.

Table 3.1 Pooling of Variance Estimates of Percent Milkfat in Packaged Milk

Set	Measurements	$\bar{x}$	SS	DF	s^2
1	2.13 2.08	2.1050	0.001250	1	0.001250
2	1.97 2.01 1.98	1.9867	0.000867	2	0.000433
3	1.91 1.91	1.9100	0	1	0
4	2.05 2.02	2.0350	0.000450	1	0.000450
5	1.95 2.09 2.01 2.04 2.03	2.0240	0.010320	4	0.002580
6	2.13 2.08	2.1050	0.001250	1	0.001250
7	2.21 1.84 2.10	2.0500	0.072200	2	0.036100

Pooled variance $= \dfrac{\Sigma\,(\text{SS})}{\Sigma\,(\text{DF})} = \dfrac{0.086337}{12} = 0.007195$. $s_p = \sqrt{0.007195} = 0.085$, (DF = 12).

We see that the sets, though similar, have different means; however all are of the same order of magnitude. We will assume here, that the chemical matrix does not vary appreciably from set to set. (An assumption of this type must be based on actual knowledge on the part of the experimenter, not merely on "wishful thinking.")

The correct way of combining the information from such a set of samples is to divide the sum of the SS values by the sum of the DF values. This gives a value called the "pooled" variance estimate s_p^2, and its square root is the pooled value of the standard deviation.

We have

$$s_p^2 = \frac{\Sigma\,(\text{SS})}{\Sigma\,(\text{DF})} = \frac{\Sigma[(\text{DF}) \cdot s^2]}{\Sigma\,(DF)} \tag{3.12}$$

This formula shows that the pooled variance is the weighted average of the individual variance estimates, the weights being the degrees of freedom. Thus an estimate with a larger number of degrees of freedom has a greater influence on the pooled value than one with a smaller number of degrees of freedom, a result consistent with our intuition.

It is also worth noting that the denominator of the pooled estimate of s^2, Σ DF, is equal to the total number of observations, 19 in our example, minus the number of sets, 7. Indeed, for the ith set, the degrees of freedom are $n_i - 1$, where n_i is the sample size for that set. By summing over all sets, we get:

$$DF_{total} = \Sigma (n_i - 1) = \Sigma n_i - k = N - k \quad (3.13)$$

where N is the total number of measurements and k the number of sets. Equation (3.13) shows that "degrees of freedom" *cannot* be *defined* by $n - 1$ (as it is in some texts). The proper formula is Eq. (3.13). This formula is often explained by stating that one degree of freedom is "lost" for each set, because it is "used up" for the estimation of the mean, $\bar{x}_i$. This explanation will become more meaningful when we discuss the "analysis of variance" technique. At any rate it is important to note that for the purpose of pooling we do not average standard deviations. However, as we shall see, the averaging of standard deviations is a common technique when dealing with control charts.

3.6 A RETURN TO WEIGHTING

1. Suppose that we have taken a series of samples from a population of mean μ and standard deviation σ. Let us assume that the number of observations in sample i is n_i and that there are p samples in all. Denote by $\bar{x}_i$ the average of the observations in the ith sample.

From Eq. (2.1) we infer that the grand average of the observations in all samples combined, which we denote by $\bar{\bar{x}}$, is

$$\bar{\bar{x}} = \frac{\Sigma n_i \cdot \bar{x}_i}{\Sigma n_i} \quad (3.14)$$

Now, according to Eq. (3.2), the variance of $\bar{x}_i$ is given by

$$\text{Var}(\bar{x}_i) = \frac{\sigma^2}{n_i} \quad (3.15)$$

If we now define the *weight* of $\bar{x}_i$, denoted as w_i, as the reciprocal of the variance:

$$w_i \equiv \frac{1}{\text{Var}(\bar{x}_i)} \quad (3.16)$$

then, combining Eqs. (3.15) and (3.16), we obtain

$$n_i = \frac{\sigma^2}{\mathrm{Var}(\bar{x}_i)} = \sigma^2 \cdot w_i \tag{3.17}$$

Introducing this expression into (3.14) we have

$$\tilde{x} = \frac{\sum w_i \sigma^2 \cdot \bar{x}_i}{\sum w_i \sigma^2}$$

and, dividing numerator and denominator by σ^2:

$$\tilde{x} = \frac{\sum w_i \bar{x}}{\sum w_i} \tag{3.18}$$

Thus, the weighted average, $\tilde{x}$, can be obtained from either Eq. (3.14) or Eq. (3.18). In fact, Eq. (3.17) shows us that the weight w_i is proportional to the number of observations n_i in the sample. This leads us to the following generalization.

2. We consider a series of measurements x_i, all of which have the same population mean μ. However, the variances of the various x_i vary, and we denote by σ_i^2 the variance of x_i. As before, we define the weight w_i, of x_i as the reciprocal of the variance σ_i^2. Thus we have

Observation	x_1	x_2	x_3	$\cdots$	x_p
Mean	μ	μ	μ	$\cdots$	μ
Variance	σ_1^2	σ_2^2	σ_3^2	$\cdots$	σ_p^2
Weight	w_1	w_2	w_3	$\cdots$	w_p

We now consider the weighted average

$$\tilde{x} = \frac{\sum w_i x_i}{\sum w_i} \tag{3.19}$$

The expected value of $\tilde{x}$ is:

$$E(\tilde{x}) = \frac{1}{\Sigma\, w_i} E[\Sigma\, w_i x_i] = \frac{1}{\Sigma\, w_i} \Sigma[E\, w_i x_i]$$

$$= \frac{1}{Ew_i} [\Sigma\, w_i E(x_i)] = \frac{1}{\Sigma\, w_i} (\Sigma\, w_i \mu)$$

$$= \frac{1}{\Sigma\, w_i} \cdot \mu \cdot \Sigma\, w_i = \mu$$

Hence,

$$E(\tilde{x}) = \mu \tag{3.20}$$

which shows that the mean of $\tilde{x}$ is the same as that of all the x_i.

It is also easily proved that

$$\text{Var}(\tilde{x}) = \frac{1}{\Sigma\, w_i} \tag{3.21}$$

Hence the *weight* of the weighted average $\tilde{x}$; that is, the reciprocal of its variance, is simply the sum of the weights of the individual x_i:

$$w_{\tilde{x}} = \Sigma\, w_i \tag{3.22}$$

It can be proved that among all the linear combinations of the x_i, for which the mean is the same as that of the original x_i, the weighted average $\tilde{x}$ has thc smallest variance [which is given by Eq. (3.21)]. Thus, in a meaningful sense, the weighted average is the "best" single linear combination of all the observations.

It is easily seen that the calculation of $\tilde{x}$, does not require that we know the w_i, but only that we know a set of quantities

$$w_1^*,\ w_2^*,\ w_3^*,\ \ldots,\ w_p^*$$

that are *proportional* to the weights $w_1, w_2, \ldots, w_p$:

$$w_i^* = K^2 w_i \tag{3.23}$$

where K^2 is any positive constant. (Weights are, by definition, positive constants.) Indeed, we readily see that

$$\tilde{x} = \frac{\Sigma\, w_i^*\, x_i}{\Sigma\, w_i^*} \tag{3.24}$$

Consider now the quantity

$$\frac{\sum w_i(x_i - \tilde{x})^2}{p - 1}$$

It can be shown that the expected value of that quantity is unity:

$$E\left[\frac{\sum w_i(x_i - \tilde{x})^2}{p - 1}\right] = 1 \tag{3.25}$$

It then follows that

$$E\left[\frac{\sum w_i^* (x_i - \tilde{x})^2}{p - 1}\right] = K^2 \tag{3.26}$$

where K^2 is the quantity defined by (3.23).

Let us now suppose that a set of x_i, of common μ, has been obtained by measurement and that the corresponding relative weights w_i^* are known. Then we can calculate $\tilde{x} = (\Sigma w_i^* x_i)/\Sigma w_i^*$ as an unbiased estimate of μ, and we can also calculate the quantity

$$K^2 = \frac{\sum w_i^*(x_i - \tilde{x})^2}{p - 1}$$

where p is the number of x_i. Then, the true weights $w_1, w_2, \ldots, w_p$ can be estimated by the formula

$$w_i = \frac{w_i^*}{K^2} \tag{3.27}$$

Observe, however, that K^2 is only a sample estimate. Therefore, the w_i will be subject to sampling error.

An example will clarify these matters. In Table 3.2, we give seven values of a measurement x, in this case the concentration of beta carotene in serum, each with its own relative weight w^*. We wish to calculate the weighted average, and estimate the standard error of this weighted average. The calculations are shown in detail. It may be worth noting that the weights need not be known with great precision: if the values of w^* were changed, but their orders of magnitude preserved, the end results would hardly change.

A final remark is in order. The use of weights, as here explained

Table 3.2 Example for the Calculation of a Weighted Average and Its Standard Error

Measurement[a] (x)	Relative weight (w^*)	$x - \bar{x}$	$w^* (x - \bar{x})^2$	$w = w^*/k^2$
.146	3	.01955	.00115	54
.140	110	.01355	.02020	1985
.142	40	.01555	.00967	722
.230	9	.10355	.09650	162
.080	80	−.04645	.17261	1443
.146	50	.01955	.01911	902
.090	10	−.03645	.01329	180
Sum	302		.33253	5448

$$\bar{x} = \frac{\Sigma\, w^* x}{\Sigma\, w^*} = \frac{38.188}{302} = .12645,\ K^2 = \frac{\Sigma\, w^* (x - \bar{x})^2}{p - 1} = \frac{.33253}{7 - 1} = .05542,$$

$$s_{\bar{x}}^2 = \frac{1}{\Sigma\, w} = \frac{1}{5448} = .0001836,\ s_{\bar{x}} = \sqrt{.0001836} = .0136.$$

[a]Beta carotene in serum, mg/liter.

is justified only when the different measurements all have the same expected value, although their standard deviations are different from each other. Thus, in Table 3.2, all seven values of x measure the same quantity, though with different precision. A weighted average of values with different expected values does not appear to have a sensible interpretation.

3.7 SUMMARY

We have discussed the concepts of the precision and accuracy of measurements. We have explained the central limit theorem and its consequence: the central role of the normal distribution. We have also shown how weights can be used when the measurements considered have different precisions.

4

Sources of Variability

4.1 SCOPE

In previous chapters we have dealt with some basic concepts and methods of statistics. We have seen that statistics deals with variability. But variability does not occur without one or more causes. In many situations it is possible to define at least two major sources of variability. This chapter deals with disentangling the effects of two common sources of random variability: within-group and between-group effects. The method that is predominantly advocated for achieving this is analysis of variance. We will discuss the assumptions that must be satisfied for the analysis of variance to yield useful results and we will present an alternative method if these assumptions are not satisfied.

4.2 WITHIN-BETWEEN ANALYSIS OF VARIANCE: CONCEPTUAL APPROACH

In Table 4.1 we have reproduced, in the first three columns, the results obtained for only one of the nine materials, namely, the ninth material, in the interlaboratory study of the determination of pentosans in pulp (see Table 1.2). This set of 21 measurements is *not* a random sample from a single population. It is rather a collection of seven sets of three measurements each, every "set" being a different laboratory. The variability observed between the 21 observations is the result of the effect of *two* sources of variability: variability between replicate measurements in any one laboratory and variability between laboratories. These sources are called *within*- and *between*-laboratory variability, respectively. We can also look at these 21 numbers as seven random samples, each one from a different population, the populations being generated by replicate measurements in each of seven laboratories. Because there may be systematic differences between the laboratories, we cannot combine all 21 measurements into a single sample.

There are a number of ways of examining a set of data of this type and we will, in fact, discuss several techniques. In this section

Table 4.1 Pentosans in Pulp; Material 9 (see Table 1.2)

Laboratory	Replicate measurements			Average ($\bar{x}$)	SS
1	17.13	16.56	16.56	16.7500	0.216600
2	16.08	16.04	16.13	16.0833	0.004067
3	16.01	15.96	16.06	16.0100	0.005000
4	16.65	16.91	16.75	16.7700	0.034400
5	15.71	15.45	15.66	15.6067	0.038067
6	15.05	14.73	15.04	14.9400	0.066200
7	18.80	18.20	18.10	18.3667	0.286667

Grand mean = $\bar{\bar{x}}$ = 16.3610, Σ SS = 0.651001, Variance of averages = 1.188327, Σ SS/Σ DF = 0.046500.
s_w^2 = Pooled within laboratory variance = 0.046500, $s_{\bar{x}}^2$ = Variance of $\bar{x}$ = 1.188327,
s_L^2 = Variance between laboratories = $s_{\bar{x}}^2 - s_w^2/3$ = 1.172946.

Summary: $\bar{\bar{x}}$ = 16.3610, s_w^2 = 0.0465, s_L^2 = 1.1729.

we explain the technique called "one way analysis of variance" or "within-between analysis of variance." While it is a classical technique, covered in both theoretical and applied statistical texts, and believed by many to be the only "theoretically sound" analysis, it is *not* the best technique, unless supplemented by additional detailed examination of the data, for reasons we will discuss in detail.

The model underlying the analysis is

$$x_{ij} = \mu_i + \varepsilon \tag{4.1}$$

where x_{ij} is the jth observation of the ith set, μ_i is the mean of the ith population, and ε is the experimental error that generates the ith population. Generally, the following additional assumptions are made:

(a) All populations have the same variance, σ_w^2.

(b) The μ_i generate themselves a population, of mean μ and variance σ_L^2.

We refer to σ_w^2 as the within-laboratory (or, more generally, within-group, or within-set) variance, and to σ_L^2 as the between-laboratory (between-group, between-set) variance. The assumption (b) can be formulated by the relation:

$$\mu_i = \mu + L_i \tag{4.2}$$

whose L_i is consequently equal to $\mu_i - \mu$, and hence represents the distance of the mean of the ith laboratory, μ_i, from the grand mean μ.

Put simply, the measurements made in the same laboratory vary from each other in terms of the variance σ_w^2; and the means of the populations vary from each other in accordance with a variance σ_L^2. It is important to note that the quantity μ_i, being the (population) mean for the ith laboratory, is therefore a constant when viewed from the perspective of this laboratory only. When viewed from the perspective of the population of *all* laboratories, it is however a random variable with variance σ_L^2.

The "internal" SS_i in the table are exactly analogous to the SS of Table 3.1. They represent the variability within-sets, for each set (laboratory). It is not necessary to list the individual DF: they are all equal to 3 − 1. Consequently, the degrees of freedom for the pooled "within-laboratory" variance is 7 × (3 − 1) = 14.

The between-laboratory variance is derived from the variance between the laboratory averages, which we have denoted by $\bar{x}_i$. To obtain a quantitative relation, we first combine Eqs. (4.1) and (4.2), which gives us

$$x_{ij} = \mu + L_i + \varepsilon \tag{4.3}$$

Taking, for a specific laboratory i, the average over all replicates, j, we obtain

$$\bar{x}_i = \mu + L_i + \bar{\varepsilon} \tag{4.4}$$

where $\bar{x}_i$ is the average of the three individual measurements obtained by laboratory i. The variance of $\bar{x}_i$ is equal to

$$\text{Var}(\bar{x}_i) = \text{Var}(L_i) + \text{Var}(\bar{\varepsilon}) = \text{Var}(L_i) + \frac{\text{Var}(\varepsilon_i)}{3}$$

Where Var(ε) is of course nothing other than σ_w^2, and $\text{Var}(L_i) = \sigma_L^2$. The relation also holds for the estimates of these variances:

$$s_{\bar{x}_i}^2 = s_L^2 + \frac{s_w^2}{3} \tag{4.5}$$

Now, for $s^2(\bar{x}_i)$ we have the estimate (see Table 4.1) 1.188327 with 7 − 1 = 6 degrees of freedom. The estimate s_w^2 is the pooled value of within-laboratory variability: $s_w^2 = 0.046500$. By subtraction we then obtain:

$$s_L^2 = s_{\bar{x}_i}^2 - \frac{s_w^2}{3} = 1.188327 - \frac{.046500}{3} = 1.172827$$

4.3 FORMAL ONE-WAY ANALYSIS OF VARIANCE

The preceding presentation was chosen in order to explain the rationale underlying the technique. The usual presentation, while not as readily understandable as the one above, is nevertheless more elegant. It is called "One-Way Analysis of Variance" and is shown in Table 4.2.

It is based on the calculation of sums, rather than of averages. We coded the data prior to carrying out the calculations by subtracting 15 from all values, and multiplying by 100. Consequently, all calculated variances must be divided by 10^4 for reconversion to the original units. The table is shown in four parts. Part A is self-explanatory. Part B uses the following general rule: In any sum of squares of the form $\Sigma_k S_k^2$, if S_k is the sum of n *individual* measurements, the *divisor* to be used for ΣS_k^2 is n. Thus, each S is the sum of three individual measurements. Therefore, the divisor for ΣS^2 is 3. T is the sum of 21 measurements; therefore T^2 has the divisor 21. (There is only one T, therefore the sum of squares of T, ΣT^2, consists only of T^2.) The sum $\Sigma (\Sigma x^2)$ is a sum of the squares x^2 of *individual* measurements; therefore the divisor is *one*.

Part C is the "Analysis of Variance Table" proper. It contains the two "sources" of variation "between" and "within," and also the source labeled "Total." This is a fictitious source; it would be the only source, if the entire collection of measurements consisted of 21 replicates from a single population. In that case there would be 20 degrees of freedom and the (coded) variance of this population would be 11,020. Because of the existence of seven populations (rather than one), there is a *partitioning* of the degrees of freedom as well as of the SS into *additive* parts. Thus, the 20 degrees of freedom for "total" are now partitioned into $6 = (7 - 1)$, for "between" and 14, for "within" (note that $6 + 14 = 20$). An entirely analogous statement holds for the SS, *but not for the MS*. In other words, the MS are *not* additive, while the DF and SS are.

The last column in C, labeled E(MS), contains the "expected values of the mean squares" of the previous column. To understand this concept, we observe that the values given in the MS column are *random variables*. If we repeated the entire experiment, one would get different values in this column, and if we repeated it a very large number of times, each of these three MS would generate a population of MS values. The last column gives the expected values of these populations as functions of σ_w^2 and σ_L^2.

Part D consists of the calculation of the estimates of σ_w^2 and σ_L^2. Previously, we have denoted these by s_w^2 and s_L^2. Here we used a different, but also commonly used notation: an estimate of a population

Table 4.2 Analysis of Variance of Coded Data (see Table 4.1)
Part A: Data and Partial Sums

Laboratory		Coded data		Sums (S_k)
1	213	156	156	525
2	108	104	113	325
3	101	96	106	303
4	165	191	175	531
5	71	45	66	182
6	5	−27	4	−18
7	380	320	310	1010

Grand sum = 2858 = $\Sigma\, S_k$

Part B: Sums of Squares and Divisors

(a) *Single Measurements (x)*

$\Sigma\, x^2 = 609{,}366$, Divisor = 1 } Sum of squares = 609,366/1 = 609,366

(b) *Sums of Data Within Laboratories* (S_k)

$\Sigma\, S_k^2 = 1{,}808{,}568$, Divisor = 3 } Sum of squares = 1,808,568/3 = 602,856

(c) *Grand sum* ($\Sigma\, S_k$)

$(\Sigma\, S_k)^2 = 8{,}168{,}164$, Divisor = 21 } Sum of squares = 8,168,164/21 = 388,960.1905

(d) *Corrected sums of squares*

Total: 609,366 − 388, 906.1905 = 220,405.8095
Laboratories: 602,856 − 388,960.1905 = 213,895.8095
Within laboratories: (total) − (laboratories) = 6,510

Part C: Analysis of Variance Table

Source	DF	SS	MS	E(MS)
Total	20	220,405.8095	11,020.2905	—
Between labor.	6	213,895.8095	35,649.3016	$\sigma_w^2 + 3\sigma_L^2$
Within labor.	14	6,510	465	σ_w^2

(continued)

Table 4.2 Continued
Part D: Components of Variance

$$\hat{\sigma}_w^2 = 465$$
$$\hat{\sigma}_w^2 + 3\hat{\sigma}_L^2 = 35{,}649$$

Hence, $\hat{\sigma}_L^2 = (35{,}649 - 465)/3 = 11{,}728$.

In original (uncoded) units,

$$\hat{\sigma}_w^2 = 465/10^4 = 0.0465$$
$$\hat{\sigma}_L^2 = 11{,}728/10^4 = 1.1728$$

---

parameter is here denoted by the same symbol as the parameter, but with a "caret" (the symbol ^) on top of it.

The estimates are obtained by "equating the MS to their expected values"; in this case from the two equations

$$35{,}649 = \hat{\sigma}_w^2 + 3\hat{\sigma}_L^2$$
$$465 = \hat{\sigma}_w^2$$

by solving for $\hat{\sigma}_w^2$ and $\hat{\sigma}_L^2$.

We now make a number of observations consisting of additional facts as well as some criticisms of the procedure we have just described. (1) The technique does not require equal numbers of replicates for each set. A set such as the one shown in Table 3.1 could readily be analyzed in an analogous way, but would require some modifications in the calculations. We will resume discussion of this topic in Section 4.3. (2) Usually, "tests of significance" are carried out on the MS, as part (often considered the most important part) of the analysis of variance. We omit these tests here, but will discuss the matter in a general way further in this book. (3) We have made no "assumptions of normality," that is, assumed that all the populations mentioned (those of the ε and that of the L_i) are "normal." What we have shown and discussed does not require such assumptions. (4) It happens occasionally that the estimate of σ_L^2, when calculated according to the procedures above, is negative. This is the result of sampling fluctuations in the mean squares and is more likely to occur when $\hat{\sigma}_L^2$ is relatively small. Of course, a variance cannot be negative, and therefore whenever the formulas give a negative value for $\hat{\sigma}_L^2$, this quantity is set equal to zero. The inference from the analysis, in that case, is that there is no noticeable variability between the population groups measured; in other words that all L_i are essentially zero. (5) Assumptions (a) and (b), which were the basis of our reasoning in our first exposition of the technique, are by no means always true. Starting with (a), it could well happen that one of the (in this case, seven) participating laboratories, works with less care and strict attention to the requirements of the chemical analytical procedure than the others. It would then be likely to have a larger σ_w^2 than the others. For our data, for example, both laboratory 1 and laboratory 7, seem to show greater variability than the five others. Is this real or just the effect of chance? The usual way to handle such a question is to perform a test of significance. No matter what such a test shows, the conclusion will always remain doubtful when it is based, as it is in this case, on variance estimates based on two degrees of freedom. We will attempt, later, to cope with this matter in what we consider a more adequate manner. Turning now to assumption (b), this too must be carefully considered. The term "to generate a population" implies a *random* process. Now, the laboratories participating in an interlaboratory study are never a "random drawing" from *all* laboratories, for many reasons, the most obvious being that participation in an interlaboratory study is a voluntary act based on special technical or commercial interests, on the availability of sufficient extra funds and of laboratory personnel, etc. As we shall see later, a good interlaboratory study requires that all volunteering laboratories be familiar with the measuring process that is being studied, that they have had previous experience with it, that they be properly equipped, reliable, with competent technical and managerial personnel, that they be aware of the importance of the study and take it very seriously. The inferences drawn from a study involving such a selection of laboratories cannot possibly apply to *all* laboratories. At best they will be valid for laboratories *similar* to those who participated in the study. Thus, the random requirement can only apply here in the sense that the participating laboratories be considered as a random sample from a *hypothetical* population of laboratories satisfying the requirements enumerated above. In most cases, this hypothetical population *does not exist at all* in reality, and is therefore either *purely* hypothetical (and imaginary), or else limited to the small set of participating laboratories. In any case, real life tells us that often one or more laboratories will show systematic effects that appear to be much larger than others. Our illustrative example is a case in point: laboratory 6 seems exceptionally *low* and laboratory 7 exceptionally *high*. Again we may ask if this is "real" or the "result of chance," and again any answer will be very tentative, because of the relatively small number of laboratories involved. Of course, if the discrepancies for laboratory 7 had been more extreme, we might have good reasons for suspecting "blunders," but at what point do systematic errors end and blunders begin? Tests of significance can be useful guides but they do not lead to final, certain verdicts. Furthermore, when more than one laboratory is suspected of being an "outlier," the tests of significance are no longer straightforward in their underlying logic.

For all these reasons, the results obtained from an analysis of variance such as the one above are not unconditionally acceptable. They must be supplemented by a careful and detailed examination of the data. In the end, the judgment of the data analyst, his knowledge of the measurement process under study, and information derived from additional data will come into play. We will have an opportunity to return to this matter.

4.4 AN ITERATIVE CALCULATION APPROACH

A new approach has been developed for the analysis of data classified within and between groups [Mandel and Paule, 1970] and [Paule and Mandel, 1989]. It is more general than the techniques presented in Sections 4.1 and 4.2, in that it allows for differing numbers of replicate measurements in the various groups, and also for different replication variances ("within-group" variances) in these groups.

We return to the model expressed by Eq. (4.3), but write it now as

$$x_{ij} = \mu + L_i + \varepsilon_{ij} \qquad i = 1, \ldots, p \tag{4.6}$$

where L_i is the *effect* (deviation from μ) of the ith group,* and ε_{ij} is

*The groups may be laboratories, but any other grouping such as days, instruments, methods, etc., is equally acceptable.

the *error* (deviation from the true group average) of the jth replicate in the ith group. The conditions postulated in Sections 4.1 and 4.2 are relaxed in two ways. First, we do not require that the number of replicate measurements be the same in all groups. Therefore, we introduce the variable n_i, which represents the number of replicates in the ith group.

Secondly, we do not require that the within-group variance, which we formerly denoted by σ_w^2, be the same for all groups. Consequently, we denote the within-group variance for the ith group by σ_i^2.

The between-group variance, which is the variance of L in Eq. (4.6), is denoted, as before, by σ_L^2.

Summing both members of Eq. (4.6) over j and dividing by n_i, we obtain

$$\bar{x}_i = \mu + L_i + \varepsilon_i \tag{4.7}$$

where

$$\bar{x}_i = \frac{\sum_j x_{ij}}{n_i} \tag{4.8}$$

$$\varepsilon_i = \frac{\sum_j \varepsilon_{ij}}{n_i} \tag{4.9}$$

From the properties of variances we presented in Chapter 2, it follows that

$$\text{Var}(\bar{x}_i) = \sigma_L^2 + \frac{\sigma_i^2}{n_i} \tag{4.10}$$

By definition, the weight w_i of $\bar{x}_i$, is given by

$$w_i = \frac{1}{\sigma_L^2 + \sigma_i^2/n_i} \tag{4.11}$$

Let

$$\tilde{x} = \frac{\sum w_i \bar{x}_i}{\sum w_i} \tag{4.12}$$

This is the weighted average of the $\bar{x}_i$. Construct the quantity:

$$\frac{\sum w_i(\bar{x}_i - \tilde{x})^2}{p - 1}$$

Since the w_i are the true weights, the expected value of this quantity is unity (see Section 3.5). Therefore, we have, apart from sampling fluctuations:

$$\frac{\sum w_i (\bar{x}_i - \bar{\bar{x}})^2}{p - 1} = 1 \tag{4.13}$$

Define the quantity F by

$$F \equiv \sum w_i (\bar{x}_i - \bar{\bar{x}})^2 - (p - 1) \tag{4.14}$$

Then, as a result of (4.13) we must have as closely as possible:

$$F = 0 \tag{4.15}$$

Now the w_i are functions of $\sigma^2{}_i$ (Eq. 4.11). For each $\sigma^2{}_i$ we readily obtain an estimate from the original measurements x_{ij}, namely

$$\hat{\sigma}_i^2 = \frac{\sum_j (x_{ij} - \bar{x}_i)^2}{n_i - 1} \tag{4.16}$$

Therefore each w_i could be estimated if we had an estimate for σ_L^2, say $\hat{\sigma}_L^2$. Representing by w_i the weight corresponding to such an estimate we have

$$\hat{w}_i = \frac{1}{\hat{\sigma}_L^2 + \hat{\sigma}_i^2/n_i} \tag{4.17}$$

We have, so far, no estimate for σ_L^2, but since this is the only unknown quantity, we propose to find a value $\hat{\sigma}_L^2$ such that when we use the corresponding $\hat{w}_i$ in Eq. (4.14), we will have $F = 0$. In other words, we attempt to find a value $\hat{\sigma}_L^2$ such that, for $\hat{w}_i$ defined by Eq. (4.17), we obtain

$$F = \sum \hat{w}_i (\bar{x}_i - \bar{\bar{x}})^2 - (p - 1) = 0 \tag{4.18}$$

The solution of this algebraic problem is achieved through an iteration process which, in practice, requires a computer. We proceed as follows:

(a) Select an initial value for $\hat{\sigma}_L^2$, say t_0.

(b) Calculate all w_i (Eq. 4.17) and F:

$$F = \sum \hat{w}_i (\bar{x}_i - \bar{\bar{x}})^2 - (p - 1)$$

(c) If F is not zero, consider a correction Δt for t, so that $t = t_0 + \Delta t$

(d) Repeat (b).

(e) Continue the process until one obtains $F = 0$. At that point, we have the best estimate for $\hat{\sigma}_L^2$.

The method for obtaining the correction Δt, at each iteration, is the Newton-Gauss approximation formula. Suppose that we have a value for $\hat{\sigma}_L^2$ that, when used in Eq. (4.18), gives a value of F that is *not* zero, say F^*. Then we must find a correction for $\hat{\sigma}_L^2$ that brings F closer to zero. Let Δ be such a correction. Calculus teaches us that for a small value of Δ, if F is the new value, we have

$$F = F^* + \frac{dF}{d\hat{\sigma}_L^2} \Delta$$

but we wish to have $F = 0$. Hence,

$$\Delta = \frac{-F^*}{dF/d\hat{\sigma}_L^2}$$

Now it can be shown that because of (4.18):

$$\frac{dF}{d\hat{\sigma}_L^2} = -\sum \hat{w}_i^2 (\bar{x}_i - \tilde{x})^2$$

Therefore,

$$\Delta = \frac{F^*}{\sum \hat{w}_i^2 (\bar{x}_i - \tilde{x})^2}$$

or

$$\Delta = \frac{\sum w_i (\bar{x}_i - \tilde{x})^2 - (p - 1)}{\sum \hat{w}_i^2 (x_i - \tilde{x})^2} \tag{4.19}$$

Having calculated Δ, we obtain a new, "corrected" value for $\hat{\sigma}_L^2$:

$$\hat{\sigma}_L^2 = \hat{\sigma}_L^2 + \Delta$$

This process is repeated until the value of F is very close to zero. Several remarks are in order.

1. As was mentioned in connection with the analysis of variance (Sections 4.1 and 4.2), it may happen that the estimate of σ_L^2 is negative. By the present method of analysis, this occurs when the solution of the algebraic equation (4.15), with F given by Eq. (4.14) is a negative value for $\hat{\sigma}_L^2$. In such cases, we set σ_L^2 equal to zero.

2. It can be shown that as a result of the properties of the first and second derivative of F with respect to $\hat{\sigma}_L^2$, the iteration process will always converge under the following conditions: (a) the true solution of Eq. (4.14) is a positive value of $\hat{\sigma}_L^2$; (b) the initial value chosen for $\hat{\sigma}_L^2$ is positive.

3. As a result of the previous observation (2), a recommended procedure is to start with a value of σ_L^2 slightly above zero. If upon iteration a negative value is obtained, this indicates that the true solution of Eq. (4.18) is negative. In that case, σ_L^2 is set equal to zero and the process is terminated.

4. In some cases it is reasonable to assume, on the basis of prior knowledge, that all σ_w^2 are equal. In such cases, all $\hat{\sigma}_i^2$ in Eq. (4.17) are replaced by their pooled value, say $\hat{\sigma}_w^2$. Apart from that modification, the computational process is the same as above.

4.5 A NUMERICAL ILLUSTRATION OF THE ITERATIVE PROCEDURE

We illustrate the procedure with the example shown in Table 4.1, but assuming that some of the laboratories tested fewer than three replicates. We also assume that we cannot consider all seven within-laboratory (population) standard deviations to be equal. The data are shown in Table 4.3. Since laboratory 4 made only a single measurement, we have no estimate for its standard deviation. Let us assume that an average of the six other standard deviations gives an acceptable estimate for laboratory 4. This value is 0.2068, which we consider to be the estimate for laboratory 4.

We must start with an initial estimate for $\hat{\sigma}_L^2$. Let it be $\hat{\sigma}_L^2 = 0.1$. Under those conditions the weights (Eq. 4.11) are given by the first column in Table 4.4. The iteration process yields successively columns 2 to 7 in Table 4.4. The corresponding values of $\hat{\sigma}_L^2$ and of F are given in the two bottom rows. At iteration 7 the value of F is very close to

Table 4.3 Hypothetical Data for Pentosans (see Table 4.1)

Laboratory	Replicate measurements			Average ($\bar{x}_i$)	Within-laboratory standard deviation (S_4)
1	17.13	16.56	—	16.845	.4030
2	16.08	16.04	16.13	16.083	.0451
3	16.01	15.96	16.06	16.010	.0500
4	16.65	—	—	16.650	.2068
5	15.71	15.45	15.66	15.607	.1380
6	15.05	14.73	—	14.890	.2263
7	18.80	18.20	18.10	18.367	.3786

zero; further iteration would leave the weights essentially unchanged. Using the weights of iteration 7, and the corresponding value of $\hat{\sigma}_L^2 = 1.174$, we can now calculate the two standard deviations, within-laboratory and between laboratories. The latter is of course the same for all laboratories and is equal to $\sqrt{1.174} = 1.083$. It may be noted that the final weights (iteration 7 in Table 4.4) differ very little from each other. The reason for this is that in this case, the between-laboratory variability is much larger than the ten quantities of $\hat{\sigma}_i^2/3$. Thus, the weight is determined predominantly by the between-laboratory variability, which is the same for all laboratories.

The calculations necessary to obtain the weights in Table 4.4 are

Table 4.4 Weights

Group	1	2	3	4	5	6	7
1	5.518	3.302	1.974	1.261	0.927	0.813	0.797
2	9.933	4.498	2.347	1.404	1.002	0.870	0.852
3	9.917	4.495	2.346	1.404	1.002	0.870	0.851
4	7.004	3.782	2.136	1.326	0.961	0.839	0.822
5	9.403	4.386	2.316	1.393	0.996	0.866	0.847
6	7.962	4.045	2.217	1.357	0.977	0.852	0.834
7	6.767	3.712	2.113	1.317	0.957	0.836	0.819
σ_L^2	0.222	0.425	0.712	0.998	1.148	1.174	1.174
F	46.852	22.035	9.700	3.705	1.024	0.130	0.003

too tedious to be performed without a computer. A computer program is easily written for these calculations, including a criterion to determine where to stop the iteration process.

4.6 CONSENSUS VALUES

The within-between model often occurs under the following circumstances. A reference material is prepared in relatively large quantity to serve either as a calibration material or as a control material for one or many laboratories. Standard agencies such as the U.S. National Bureau of Standards, or the National Physical Laboratory in Great Britain issue many such materials to assist laboratories in their country or abroad in calibrating their measuring equipment or to monitor a measuring process over a period of time. It is then necessary to determine as accurately as possible the pertinent properties of this material. An example would be the manganese content of a standard steel.

The standard material is prepared, thoroughly homogenized so that all portions of it are virtually identical, and then measured, say for its manganese content. Often several laboratories are involved in this measuring process. It is also common to perform the measurement by two or more different analytical techniques. Each laboratory may make several replicate measurements, the number often varying from laboratory to laboratory. The same is true for the different methods of test. Thus, one obtains data from various sources (laboratories, methods) with unequal replication for the different sources.

If one has no reason to rank one technique over another, or one laboratory over another, then one must accept the values originating from all sources and derive a "consensus value" from the data.

One can use the technique presented in the previous section for the purpose of calculating a consensus value. The consensus value would then be given by $\tilde{x}$, as given by Eq. (4.12). The w_i used in this equation would be the estimates $\hat{w}_i$ given by Eq. (4.17), with $\hat{\sigma}_L^2$ obtained from the iterative calculation process.

The question arises, what precision can one associate with such an $\tilde{x}$ value? The answer is provided by Eq. (3.22):

$$w_{\tilde{x}} = \sum_i w_i$$

Hence

$$\hat{\sigma}_{\bar{x}}^2 = \frac{1}{\sum_i w_i} \tag{4.20}$$

and

$$\hat{\sigma}_{\bar{x}} = \frac{1}{\sqrt{\sum_i w_i}} \tag{4.21}$$

It is true that the w_i are actually only sample estimates of the true w_i, resulting in perhaps considerable uncertainty in $\hat{\sigma}_{\bar{x}}$. The larger the experiment, the smaller will be this uncertainty, and one is therefore, as usually, confronted with the problem of balancing cost versus amount of information provided by the experiment. Fortunately it is seldom necessary to know $\sigma_{\bar{x}}$ with great accuracy, and the value provided by the experiment will often be quite sufficiently accurate for the practical purposes intended.

4.7 SUMMARY

In this chapter we have become familiar with the analysis of variance technique for measuring within- and between-group variability. We have seen that the usual assumptions of analysis of variance are not always fulfilled, and we have presented an alternate, more general technique that allows for more flexible assumptions.

REFERENCES

Mandel, J., and R. C. Paule (1970). Interlaboratory Evaluation of a Material with Unequal Number of Replicates. Anal. Chem., *42*, 1194–1197.

Paule, R. C., and J. Mandel (1982). Consensus Values and Weighting Factors, J. Res. Natl. Bureau of Standards, *87*, 377–385.

Paule, R. C., and J. Mandel (1989). Consensus Values, Regressions, and Weighting Factors, J. Res. Natl. Institute of Standards and Technology, *94*, 197–203.

5

Linear Functions of a Single Variable

5.1 SCOPE

An important situation in science is that in which a straight line has to be fitted to a set of pairs of values x,y. Many calibration curves are linear and require a proper procedure for the exact calculation of the straight line parameters. In this chapter we cover three common cases: the classical case, in which x is without error, and the error variance of y is constant over the entire curve; the weighted case, in which the variance of the error of y varies in a known relative fashion; and the so-called "error-in-variable" case, in which both x and y are subject to error, each with its own constant standard deviation of error. Here we can obtain a solution even when the error variances are unknown, as long as the ratio of their variances is known.

5.2 FITTING STRAIGHT LINES TO DATA: THE CLASSICAL CASE

The subject of this section is often referred to as "linear regression," a name which is still used predominantly in the statistical literature, but has only historical significance in terms of the meaning of its words.

The problems we are about to discuss are best presented in terms of a numerical example. Table 5.1 lists, in the columns labeled x and y, the concentration of glucose in seven samples of serum, and the corresponding "absorbance" measurements on these samples, using a spectrophotometric method. Once the relation between x and y has been established, it can be used as a means of determining the glucose content of an "unknown" sample of serum, by measuring the absorbance by the same method and under similar conditions as were used in the experiment leading to the data listed here. Thus these data constitute a "calibration" experiment, to calibrate the spectrophotometric measurement in terms of glucose concentration.

Suppose that we can find a relation of the type

$$y = \alpha + \beta x \tag{5.1}$$

between x and y. Then, in the future, for any given sample of glucose, if we measure its absorbance y, we can calculate x by the relation

$$x = \frac{y - \alpha}{\beta}$$

Table 5.1 Glucose in Serum Calibration Experiment

Concentration of glucose mg/dl x	Absorbance y
0	0.050
50	0.189
100	0.326
150	0.467
200	0.605
400	1.156
600	1.704

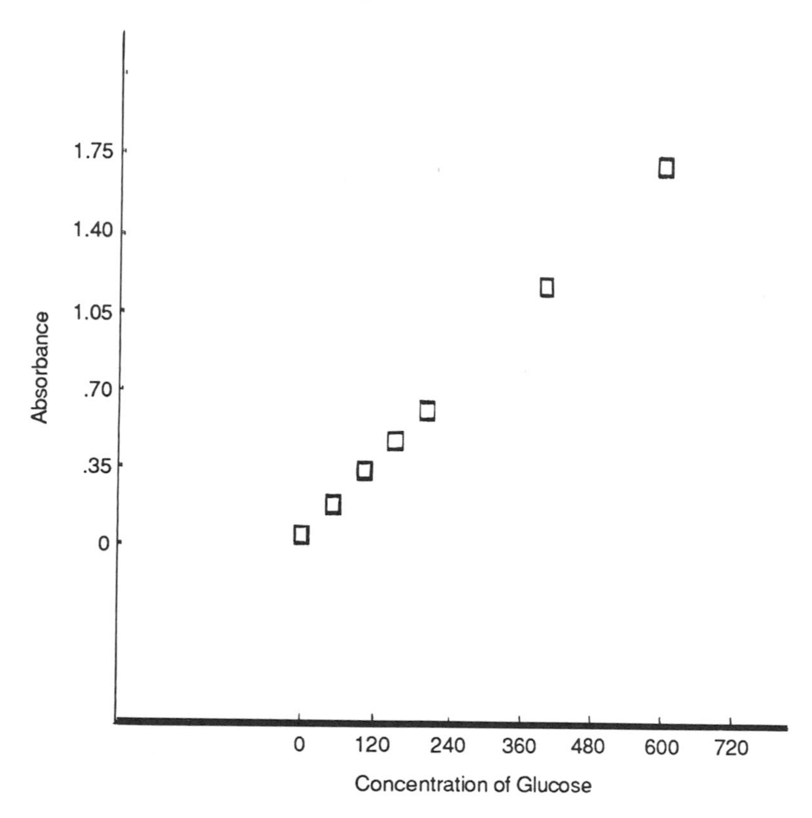

Figure 5.1 Calibration of glucose analysis.

the values of α and β having been determined by the calibration experiment.

A plot of y versus x, shown in Figure 5.1, reveals however that the points do not lie *exactly* on a straight line. This is due to experimental error in the y-measurements, since the x are known with only negligible error. Therefore, we rewrite Eq. (5.1) more exactly as:

$$y = \alpha + \beta x + \varepsilon \tag{5.2}$$

where ε is an experimental error in the y-measurement.

We will assume, in this section, that the ε all belong to a single population of errors, of mean zero, and standard deviation σ. The assumption of zero mean is plausible since on the average the positive errors will tend to cancel the negative ones. At any rate, if their average is not zero it can be incorporated in α. The standard deviation σ is not known, but can as we shall see, be estimated from the calibration experiment. The expression "fitting a straight line to the data" means to find appropriate values for α and β which are, of course, respectively the *y-intercept* and the *slope* of the straight line.

Suppose that we have found such a pair of values, which we denote by $\hat{\alpha}$ and $\hat{\beta}$. Then for each x, we can calculate a "fitted y" by the relation

$$\hat{y} = \hat{\alpha} + \hat{\beta}x \tag{5.3}$$

and compare this fitted value with the *observed* value y corresponding to the same value of x. The difference

$$d = y - \hat{y} \tag{5.4}$$

is called the *residual*.

Equation (5.3) is the equation of a straight line. All $(x,\hat{y})$ points lie *exactly* on the fitted line. The points (x,y), on the other hand, will scatter about this line. The vertical distance between y and $\hat{y}$, at a given x, is precisely the residual d of Eq. (5.4). The object of the fitting process is to fit the line so as to:

(a) balance the positive and negative d;

(b) make the d as small as possible.

This is accomplished by the "method of least squares" according to which the following two conditions are fulfilled:

(a) $\Sigma d = 0$, where the summation is over all experimental points;

(b) Σd^2 is as small as it can be made.

Condition (b) justifies the name "least squares," which is short for "smallest possible sum of squares of residuals."

By elementary calculus, the least squares requirement leads to the following way of determining $\hat{\alpha}$ and $\hat{\beta}$.

We first calculate the averages $\bar{x}$ and $\bar{y}$ and the following three quantities:

$$S_{xx} = \Sigma\,(x - \bar{x})^2 \tag{5.5a}$$

$$S_{yy} = \Sigma\,(y - \bar{y})^2 \tag{5.5b}$$

$$S_{xy} = \Sigma\,(x - \bar{x})(y - \bar{y}) \tag{5.5c}$$

The first two we have already encountered as the "sum of squares" of x and y, respectively. The third is the sum of the cross-products of the deviations of x and y from their respective averages $\bar{x}$ and $\bar{y}$.

We now calculate $\hat{\beta}$ and $\hat{\alpha}$ as follows:

$$\hat{\beta} = S_{xy}/S_{xx} \tag{5.6a}$$

$$\hat{\alpha} = \bar{y} - \hat{\beta}\bar{x} \tag{5.6b}$$

The residual at each point x is now obtained from Eq. (5.4), using Eq. (5.3) to calculate $\hat{y}$.

Finally, an estimate of σ^2, the variance of errors ε, is obtained by the formula:

$$s_\varepsilon^2 = \hat{\sigma}_\varepsilon^2 = \frac{\Sigma_i\, d_i^2}{N - 2} \tag{5.7}$$

where d_i is the ith residual and N is the number of x,y pairs in the calibration experiment. The degrees of freedom are $N - 2$, because two degrees of freedom are "used up" for the estimation of $\hat{\alpha}$ and $\hat{\beta}$. All results are shown in Table 5.2. The numerator $\Sigma\, d_i^2$ of Eq. (5.7) can be calculated directly by the formula:

$$\Sigma\, d_i^2 = S_{yy} - \frac{(S_{xy})^2}{S_{xx}} \tag{5.8}$$

but the use of this equation is not recommended, for two reasons. In the first place, $\Sigma\, d_i^2$ is calculated by Eq. (5.8) as the generally small difference of two large numbers S_{yy} and $(S_{xy})^2/S_{xx}$, and is therefore subject to serious rounding errors, unless both quantities are computed with many significant figures. Secondly, it is highly advisable to cal-

Table 5.2 Linear Fit of Data of Table 5.1

Data		Fit	Residual
x	y	$\hat{y}$	d
0	0.050	0.0516	−0.0016
50	0.189	0.1895	−0.0005
100	0.326	0.3273	−0.0013
150	0.467	0.4652	0.0019
200	0.605	0.6030	0.0020
400	1.156	1.1544	0.0016
600	1.704	1.7058	−0.0018

$\bar{x} = 214.2857$, $\bar{y} = 0.642429$, $S_{xx} = 273{,}571$, $S_{yy} = 2.079562$, $S_{xy} = 754.2571$, $S_{\hat{\beta}} = 3.62 * 10^{-6}$.
$\hat{\beta} = S_{xy}/S_{xx} = 0.002757$, $\hat{\alpha} = \bar{y} - \hat{\beta} \cdot \bar{x} = 0.0516$, $s^2 = \Sigma\, d^2/(N - 2) = 0.000,003582$, $s = 0.0019$, $s_{\bar{y}} = .00072$.

culate and display all individual residuals d_i as shown in Table 5.2. Since the x are ordered increasingly, the succession of signs of the d_i is often an indication of whether a straight line is really an adequate fit. In our illustration, the signs are: − − − + + + −. Since a negative sign indicates a y *below* the fitted line and a positive sign, a y value *above* the line, there is some indication here of *curvature,* but the evidence for curvature is not entirely conclusive in this case.

The estimates $\hat{\alpha}$ and $\hat{\beta}$ are not "statistically independent." To understand this statement, we recall that both $\hat{\alpha}$ and $\hat{\beta}$ are only sample estimates of the population parameters α and β, and differ from these population parameters by "sampling errors." Thus, we may write

$$\hat{\alpha} = \alpha + e_{\hat{\alpha}} \tag{5.8a}$$

$$\hat{\beta} = \beta + e_{\hat{\beta}} \tag{5.8b}$$

The estimates $\hat{\alpha}$ and $\hat{\beta}$ can each be considered as a member of a population generated by repeating the entire straight-line experiment an unlimited number of times. Thus, we obtain two series:

$$\hat{\alpha}_1 \quad \hat{\alpha}_1 \quad \hat{\alpha}_3 \ \ldots$$
$$\hat{\beta}_1 \quad \hat{\beta}_2 \quad \hat{\beta}_3 \ \ldots$$

and the corresponding series of errors:

$$(e_{\hat{\alpha}})_1 \quad (e_{\hat{\alpha}})_2 \quad (e_{\hat{\alpha}})_3 \ldots$$

$$(e_{\hat{\beta}})_1 \quad (e_{\hat{\beta}})_2 \quad (e_{\hat{\beta}})_3 \ldots$$

Now this latter pair of series will show a negative correlation: for a *large* value of $e_{\hat{\alpha}}$ these will tend to correspond a *low* value of $e_{\hat{\beta}}$, and vice versa.

For this reason, it is often advisable to write the equation of a straight line in terms of a different pair of parameters: slope and the "height at the centroid." If $\bar{x}$ is the average of the x-values and $\bar{y}$ the average of the y-values, then the equation can be written as

$$y_i = \bar{y} + \hat{\beta}(x_i - \bar{x}) + d_i \tag{5.9}$$

It can be shown that $\bar{y}$ is precisely the ordinate of the fitted line when the abscissa is $\bar{x}$. Thus, the point $(\bar{x},\bar{y})$ is a point on the fitted line. Now, the two parameters are $\bar{y}$ and $\hat{\beta}$ and it can be shown that, for normal populations of experimental errors ε_i, $\bar{y}$, and $\hat{\beta}$ are statistically independent.

Equation (5.9) can be written as

$$y_i = (\bar{y} - \hat{\beta}\bar{x}) + \hat{\beta}x_i + d_i$$

which shows that

$$\hat{\alpha} = \bar{y} - \hat{\beta}\bar{x} \tag{5.10}$$

The standard deviations of the random variables $\bar{y}$ and $\hat{\beta}$ are given by the equations

$$\sigma_{\bar{y}} = \frac{\sigma_\varepsilon}{\sqrt{N}} \qquad \sigma_{\hat{\beta}} = \frac{\sigma_\varepsilon}{\sqrt{S_{xx}}} \tag{5.11a}$$

From Eqs. (5.10) and (5.11a) it follows that

$$\sigma_{\hat{\alpha}} = \sigma_\varepsilon \sqrt{\frac{1}{N} + \frac{\bar{x}^2}{S_{xx}}} \tag{5.11b}$$

These equations also apply when the sample estimates $s_{\bar{y}}$, $s_{\hat{\beta}}$, $s_{\hat{\alpha}}$, and s_ε are used instead of the population parameters $\sigma_{\bar{y}}$, $\sigma_{\hat{\beta}}$, $\sigma_{\hat{\alpha}}$, and σ_ε.

5.3 PREDICTION IN THE CLASSICAL CASE

Once a calibration line has been established, it is used to convert a measured y value into an x value. Thus, referring to Table 5.1, we may wish to find the glucose content of a "new" sample, after measuring its absorbance. As mentioned before, this is accomplished by solving the regression equation

$$y = \bar{y} + \hat{\beta}(x - \bar{x})$$

for x.

Denoting by y_n the measured absorbance of the new sample, and by $\hat{x}$ its calculated glucose concentration, we have

$$\hat{x} = \bar{x} + \frac{y_n - \bar{y}}{\hat{\beta}} \tag{5.12}$$

Thus, if in connection with the calibration line obtained in Table 5.1, we made a measurement on an unknown new sample and obtained an absorbance of say, 0.147, we would have (see Table 5.2):

$$\hat{x} = 214.2857 + \frac{0.147 - 0.6424}{.002757} = 34.60$$

The question arises: what is the uncertainty of this value? A simple way is to apply the law of propagation of errors (see Chapter 2). We have

$$\frac{\mathrm{Var}(\hat{x})}{\hat{x}^2} = \frac{\mathrm{Var}(y_n - \bar{y})}{(y_n - \bar{y})^2} + \frac{\mathrm{Var}(\hat{\beta})}{\hat{\beta}^2}$$

Now, y_n and $\bar{y}$ are independent, since y_n was measured after completion of the calibration experiment. Hence

$$\mathrm{Var}(y_n - \bar{y}) = \mathrm{Var}(y_n) + \mathrm{Var}(\bar{y}) = \sigma_\varepsilon^2 + \frac{\sigma_\varepsilon^2}{N}$$

$$= \sigma_\varepsilon^2\left(1 + \frac{1}{N}\right)$$

We also know that

$$\mathrm{Var}(\hat{\beta}) = \frac{\sigma_\varepsilon^2}{S_{xx}}$$

Thus, we obtain

$$\mathrm{Var}(\hat{x}) = \hat{x}^2 \left[\frac{1 + \frac{1}{N}}{(y_n - \bar{y})^2} + \frac{1}{\hat{\beta}^2 S_{xx}} \right] \sigma_\varepsilon^2 \qquad (5.13)$$

Since σ_ε^L has been estimated by s_ε^2, we substitute this value for σ_ε^L. Applying this formula to our case, with $y_n = 0.147$, we have

$$\mathrm{Var}(\hat{x}) = (34.60)^2 \left[\frac{1 + 1/7}{(.147 - .6425)^2} + \frac{1}{(.002757)^2\,(273571)} \right] (.0019)^2 = .02217$$

Hence, $\hat{\sigma}_{\hat{x}} = .15$.

Equations (5.12) and (5.13) can be recorded on the computer, or even on a simple, programmable hand calculator, with the proper numerical values and with y_n as the independent input variable. In this way, $\hat{x}$ values and their standard errors can be obtained in a matter of seconds for any given y_n.

5.4 WEIGHTED STRAIGHT LINE FITTING

Situations frequently occur in which the assumption that all ε belong to a single statistical population is not justified. Then the procedure given above is not the best one available and may indeed lead to erroneous interpretations of the data. The data in Table 5.3 represent measurements of the "strain" of eight samples of rubber. The x values are the averages of measurements for each rubber made in 15 different laboratories. We can be confident that the errors of these averages are small enough to be ignored, when compared to those of the y, which are single measurements made in only one laboratory. It is evident, from a mere inspection of the data that the y-value, 194.9, obtained for sample number 4, is not consistent with the corresponding x-value,

Table 5.3 Strain of Natural Rubber

	Data			Fit	Residual		
Material	x	y	w	$\hat{y}$	d	$\hat{w}_a$	s
1	40.3	36.9	4.12	35.52	1.38	0.0331	5.5
2	63.8	57.9	16.61	59.60	−1.70	0.1336	2.7
3	75.7	76.0	7.82	71.80	4.20	0.0629	4.0
4	99.6	194.9	0.03	96.29	98.61	0.0002	70.7
5	106.4	106.0	1.02	103.26	2.74	0.0082	11.0
6	108.1	94.4	1.69	105.00	−10.60	0.0136	8.6
7	140.4	138.0	56.14	138.10	−0.10	0.4516	1.5
8	160.8	166.9	1.00	159.01	7.89	0.0080	11.2

$\bar{x} = 114.83$, $\bar{y} = 111.90$, $S_{xx} = 117{,}087.51$, $S_{yy} = 123{,}716.12$, $S_{xy} = 119{,}992.64$.
$\hat{\beta} = S_{xy}/S_{xx} = 1.0248$, $\hat{\alpha} = \bar{y} - \hat{\beta} \cdot \bar{x} = -5.78$.
$\hat{K}^2 = 745.37/(8 - 2) = 124.31 \rightarrow \hat{K} = 11.5$, $\hat{w}_a = w/\hat{K}^2$; $s = 1/\sqrt{\hat{w}_a}$.

99.6, in comparison to the way the remaining y-date agree with the corresponding x-values. It is known that material 4 was especially sensitive to ambient humidity conditions, and that the laboratory in which the y-measurements were made was not in good control of its relative humidity. We could of course choose to omit the point (99.6, 194.9) from the fitting procedure. However, there is evidence, not shown here, that the other rubbers also showed various degrees of sensitivity to the laboratory environment, and that these could be quantitatively expressed by the "relative weights" shown in the third column of the table. The *weight* of a measurement is defined, as in Section 3.5, as the inverse of the variance of its experimental error, σ^2. Thus if a measurement y is equal to

$$y = E(y) + \varepsilon \tag{5.14}$$

where ε is the experimental error of y and $E(y)$ its expected value, then the absolute weight of y, denoted by $w_a(y)$, is given by

$$w_a(y) = \frac{1}{\sigma_\varepsilon^2} \tag{5.15}$$

For a set of N y-values, each with a different variance, there will be a set of associated weights $(w_a)_1$, $(w_a)_2$, . . . , $(w_a)_N$, defined in accordance with Eq. (5.15) as the reciprocals of the error variances.

In practice, the variances $\sigma^L_{\varepsilon i}$ and consequently also the weights $(w_a)_i$ are often unknown. It happens however frequently that the *ratios* of these weights relative to each other can be inferred from either theoretical considerations or additional data. In such cases we do have a set of positive numbers, the *relative weights,* which we denote here by w_i, that are *proportional* to the $(w_a)_i$. We then have

$$w_i = K^2(w_a)_i \tag{5.16}$$

where the constant of proportionality K^2 is not known.

Let us now consider a new variable Z_i, is defined by

$$Z_i = y_i \sqrt{w_i} \tag{5.17}$$

Then, if we denote the error of Z_i by δ_i, the standard deviation of δ_i is

$$\sigma_{\delta_i} = (\sigma_{\varepsilon_i}) \sqrt{w_i}$$

which, because of Eq. (5.15), becomes

$$\sigma_{\delta_i} = \frac{1}{\sqrt{(w_a)_i}} \sqrt{w_i}$$

and, as a result of Eq. (5.16):

$$\sigma_{\delta_i} = \frac{1}{\sqrt{(w_a)_i}} K \sqrt{(w_a)_i} = K \tag{5.18}$$

Thus, the variable Z_i has an error variance independent of *i:* all Z_i in the series have the same error variance, equal to K^2. Such a set is called *homoscedastic*. A set that is not homoscedastic, such as the y_i, is called *heteroscedastic*.

The set of Z_i values can of course be numerically calculated from Eq. (5.17), since the w_i are known. It is also important to note that the set of w_i can be replaced by any proportional set of numbers. This yields a different set of Z_i values, but the latter will still be homoscedastic. In Table 5.3, the w_i have been calculated so as to make the relative weight of sample number 8 equal to unity.

It is seen that due to their varying sensitivities to surrounding conditions, the eight rubbers had appreciably different weights. Rubber number 4 is especially noteworthy, because of this exceptionally low relative weight (high sensitivity to test conditions and consequently large experimental error).

The method of least squares is still applicable but requires some modifications in the formulas, when weights are considered. One can envision the procedure as being carried out on the homoscedastic z-transforms, and then converted back to the y scale. This amounts to minimizing $\Sigma\, w_i\, d_i^2$, rather than $\Sigma\, d_i^2$. It will then turn out that $\Sigma\, w_i d_i = 0$, instead of the relation $\Sigma\, d_i = 0$. The quantity $\Sigma\, w_i\, d_i^2$ is called the "weighted sum of squares of residuals."

Implementing these changes, we obtain the following solution, called *weighted least squares:*

1. Instead of the averages $\bar{x}$ and $\bar{y}$, calculate the "weighted averages" $\tilde{x}$ and $\tilde{y}$, defined as:

$$\tilde{x} = \frac{\Sigma\, w_i x_i}{\Sigma\, w_i} \qquad \tilde{y} = \frac{\Sigma\, w_i\, y_i}{\Sigma\, w_i} \tag{5.19}$$

2. Instead of Eq. (5.5), compute the following "weighted" quantities:

$$S_{xx} = \Sigma\, w_i\, (x_i - \tilde{x})^2 \tag{5.20a}$$

$$S_{yy} = \Sigma\, w_i\, (y_i - \tilde{y})^2 \tag{5.20b}$$

$$S_{xy} = \Sigma\, w_i\, (x_i - \tilde{x})\,(y_i - \tilde{y}) \tag{5.20c}$$

3. Compute $\hat{\beta}$ and $\hat{\alpha}$ from:

$$\hat{\beta} = S_{xy}/S_{xx} \tag{5.21}$$

$$\hat{\alpha} = \tilde{y} - \hat{\beta}\tilde{x} \tag{5.22}$$

4. The variance of the error ε corresponding to any y is obtained from Eqs. (5.15) and (5.16)

$$\sigma^2_{\varepsilon_i} = \frac{1}{(w_a)_i} = \frac{K^2}{w_i} \tag{5.23}$$

An estimate of K^2, denoted $\hat{K}^2$ (see Eq. (3.26)), is given by

$$\hat{K}^2 = \frac{\Sigma\, w_i(y_i - \hat{y}_i)^2}{N - 2} = \frac{\Sigma\, w_i d_i^2}{N - 2} \tag{5.24}$$

A simple interpretation of K is derived from Eq. (5.23): K^2 is the variance of error whose relative weight is unity. As shown in Section 3.5, we can calculate estimates for the absolute weights, by dividing the relative weights by $\hat{K}^2$. This, in turn, allows us to estimate the standard deviation of each y, as the square root of the reciprocal of the absolute weight.

The calculations and the results are shown in Table 5.3.

In the example we have just discussed, the weights, though different, do not appear to be related to the magnitude of the measured quantity. They are probably more a function of the physical and chemical matrix of the individual materials than of their strain value. There are cases, however, in which the error variance of a measured quantity will be monotonically related to its magnitude. When this occurs, the error variance generally increases as the measured value increases. The usual recommendation is to deal with such situations by a *transformation of scale* of either x or y or of both x and y.

We do not advocate this procedure in general, for a number of reasons, one of them being that weighting is a more natural, and generally an entirely satisfactory, solution to the problem. Suppose that we have a situation in which

$$y = \alpha + \beta x + \varepsilon \tag{5.25}$$

and

$$\sigma_\varepsilon = k(x)^{0.6} \tag{5.26}$$

The weight for each ε is $1/\text{Var}(\varepsilon) = (1/k^2)\, x^{-1.2}$. We can take $x^{-1.2}$ as a relative weight. We use the formulas developed above, writing for w_i the numerical value of $x_i^{-1.2}$. An example is provided by Table 5.4 which deals with the determination of the bromine number of petroleum. Here x is an average of duplicate measurements in a number of laboratories and y is the average of duplicate measurements in a single laboratory. From calculations based on the within laboratory replication, it was inferred that the replication error conforms to the relation given by Eq. (5.26). The column labeled w gives the values of $x_i^{-1.2}$. The weighted analysis yields the results shown in the table.

Table 5.4 Bromine Number of Petroleum

Material	Data					
	x	y	w_a	$\hat{y}$	d	s
1	0.76	0.68	1.39	0.680	0.0005	0.14
2	1.22	1.40	0.79	1.16	0.240	0.19
3	2.15	1.75	0.40	2.132	−0.381	0.26
4	3.64	3.50	0.21	3.688	−0.188	0.36
5	10.90	11.35	0.057	11.270	0.080	0.70
6	48.21	48.50	0.0096	50.240	−1.740	1.71
7	65.39	70.20	0.0066	68.184	2.016	2.06
8	114.18	119.45	0.0034	119.143	0.307	2.87

$\bar{x} = 1.9367$, $\bar{y} = 1.9086$, $S_{xx} = 97.45762$, $S_{yy} = 106.4852$, $S_{xy} = 101.7913$.

$\hat{\beta} = \dfrac{S_{xy}}{S_{xx}} = 1.0444$, $\hat{\alpha} = \bar{y} - \hat{\beta}\bar{x} = -0.1142$.

$$\hat{k}^2 = \frac{\Sigma\, wd^2}{N-2} = \frac{.16743}{6} = .02790,\ \hat{k} = \sqrt{.0279} = .167.$$

$$w_a = (x^{0.6})^{-2}$$

The following facts are to be noted, applying to this example and in part to weighting in general.

1. Small changes in the weights have little effect on the fit; therefore the weights need not be given with many decimals. (We could have rounded the weights in the column w to two significant figures without changing the results.)

2. The importance of weighting resides not so much in getting the best values for the slope and the intercept. These are relatively insensitive to the exact weights. The real point in using a weighted analysis is that it allows us to make proper inferences about the *uncertainty* of values derived from the fit. In particular, we may ask what the scatter of experimental points will be at various values of x. Without weighting, this scatter is assumed to be uniform over the entire range of x-values. With weighting, it obeys the weighting equation (Eq. (5.26) for our example.) The analysis will give us an estimate for the constant k, so that σ_ε can be estimated for our example from the equation:

$$s_\varepsilon = \hat{k}(x)^{0.6} \tag{5.27}$$

Our weighted analysis gives an estimate of $\hat{k} = 0.167$, which in turn provides the estimates of s_i shown in the table according to the formula given by Eq. (5.26). It is interesting to compare these results with those yielded by an unweighted analysis, by which we would have obtained the estimates $\alpha = -0.2554$; $\hat{\beta} = 1.0504$; $s_\varepsilon = 1.079$. It is the latter value that we ought to examine carefully. When applied to the samples with low bromine number, this value of uncertainty is unreasonably high: when determining the bromine number of a sample having a bromine number less than one, by the technique used for these data, we do not expect to make an experimental error of standard deviation greater than one unit. If that were the case, the y values in the table would be surprisingly close to the corresponding x at low values of x.

By contrast, the weighted analysis based on the prior knowledge that the error standard deviation of y is proportional to $x^{0.6}$, leads to s-values that exhibit the desired trend. They reflect the considerably smaller errors in y at the low end and when compared to those at the high end of the range of y-values. Thus, the weighted fit, when needed, is far more acceptable as a faithful reflection of reality than the unweighted one.

The importance of weighting has often been (and still often is) misunderstood and misrepresented, and experimenters who use weighting are often surprised to find that the intercept and slope obtained by a weighted analysis differ little from those obtained by an unweighted analysis. It is only when the analysis is extended to an evaluation of the error variances, as was done in our illustrative examples, that the true benefit of a weighted analysis is achieved.

5.5 STRAIGHT LINE FITTING WITH ERRORS IN BOTH VARIABLES

The assumption, made so far, that x is not subject to experimental error, is not always fulfilled. We will now present the case where both x and y are subject to experimental error, but will limit our discussion to situations conforming to the following assumptions:

(a) Each x_i is affected by an error δ_i:

$$x_i = E(x_i) + \delta_i \tag{5.28}$$

and each y_i is affected by an error ε_i

$$y_i = E(y_i) + \varepsilon_i \tag{5.29}$$

(b) Both δ and ε are homoscedastic variables; in other words, the variance of δ_i is the same for all i, and the variance of ε_i is the same for all i.

(c) The errors δ and ε are mutually statistically independent.

(d) The variance of δ_i, represented by Var(δ), and the variance of ε_i, represented by Var(ε), are both unknown, but their ratio

$$\lambda = \frac{\text{Var}(\varepsilon)}{\text{Var}(\delta)} \tag{5.30}$$

is known.

We also note that Eqs. (5.28) and (5.29) imply that the expected value of each δ_i and of each ε_i is zero.

The formulas for fitting a straight line to a set of points (x_i, y_i), for $i = 1$ to N, under those conditions are somewhat different from those applying to the "classical" case, given in Section 5.1. We still make use of $\bar{x}$, $\bar{y}$, and of S_{xx}, S_{yy}, S_{xy} as given by Eqs. (5.5), but calculate the slope estimate, $\hat{\beta}$, by the following quadratic equation

$$\hat{\beta} = \frac{S_{yy} - \lambda S_{xx} + \sqrt{(S_{yy} - \lambda S_{xx})^2 + 4\lambda\, S_{xy}^2}}{2\, S_{xy}} \tag{5.31}$$

The estimate of α is given, as before, by

$$\hat{\alpha} = \bar{y} - \hat{\beta}\bar{x} \tag{5.32}$$

Furthermore, we obtain estimates for the unknown Var(δ) and Var(ε) by the following formulas. We first define d_i by

$$d_i = y_i - \hat{\alpha} - \hat{\beta}x_i \tag{5.33}$$

Let us denote the quantity $\Sigma\, d^2/(N - 2)$ by D^2:

$$D^2 = \frac{\Sigma\, d^2}{N - 2} \tag{5.34}$$

Also, let us define two quantities P and Q, as follows:

$$P = \lambda + \hat{\beta}^2$$

$$Q = \lambda^2 S_{xx} + 2\lambda\, \hat{\beta} S_{xy} + \hat{\beta}^2 S_{yy}$$

Then we have the following formulas:

For the estimates of σ_δ and σ_ε:

$$s_\delta = \frac{D}{\sqrt{P}} \qquad s_\varepsilon = \frac{D\sqrt{\lambda}}{\sqrt{P}} \tag{5.35}$$

For the estimates of σ_α and σ_β:

$$s_{\hat{\alpha}} = D\sqrt{\frac{1}{N} + \frac{\bar{x}^2 P^2}{Q}} \tag{5.36}$$

$$s_{\hat{\beta}} = D\,\frac{P}{\sqrt{Q}} \tag{5.37}$$

Finally, estimates for x_i and y_i, corrected for error, which we denote by $\hat{x}_i$ and $\hat{y}_i$, are given by the equations:

$$\hat{x}_i = x_i + \frac{\hat{\beta}}{P}\, d_i \tag{5.38}$$

$$\hat{y}_i = y_i - \frac{\lambda}{P}\, d_i \tag{5.39}$$

An example of a situation where both variables are subject to error is shown in Table 5.5, dealing with the determination of nickel in steel (Hicho, 1983). The measurement is obtained by a new method, denoted by y, and is compared to a well-established method, denoted by x. Table 5.5 shows the results on 13 samples. The values at the bottom of the table are the standard deviations for each method obtained from replicated measurements. Thus the value for λ is equal to $(.12/.005)^2 = 576$. Applying Eqs. (5.33)–(5.37), we obtain the values shown in

Table 5.5 Calibration Curve for Nickel in Steel

Sample	x	y
1	0.6078	2.112
2	0.6338	1.948
3	0.6514	2.266
4	0.6578	1.950
5	0.6746	2.045
6	0.6772	2.088
7	0.7023	2.540
8	0.7103	2.512
9	0.7300	2.330
10	0.7500	2.258
11	0.7724	2.472
12	0.8021	2.795
13	0.8185	2.954
Standard deviation of error	0.0050	0.12

$\bar{x} = 0.7068, \bar{y} = 2.3285, S_{xx} = 0.050812, S_{yy} = 1.18988, S_{xy} = .207888.$
$P = 593.1221,\ Q = 17869.7.$
$\lambda = \left(\frac{0.12}{0.005}\right)^2,\ a = -0.60,\ b = 4.14,\ S_d = .0072,\ S_e = .17,$
$S_a = .55,\ S_b = .78.$

the table. The important point is that the procedure we used for these data allow us to calculate estimates for both s_δ and s_ε. The ordinary least squares treatment would have thrown all the error in s_ε, on the assumption that $s_\delta = 0$. This would have constituted the use of an erroneous model for the data.

The following observation is similar to one we made in discussing weighted regression. The estimates $\hat{\beta}$ and $\hat{\alpha}$ will, in general not be appreciable different from those obtained by using the formulas for the "classical" case. The important feature justifying different formulas here lies in the evaluation of the *errors of measurement,* δ and ε. The modified approach, unlike the "classical" one, recognizes that both x and y are subject to error, and evaluates both sets of errors. Another interesting difference between the two cases is that, unlike the classical case, the one discussed here is *entirely symmetrical* with respect to x

and y. This means that we are at liberty to call either of the two variables x, and the other y. The end results will not be affected by this choice. Of course, if we interchange the meaning of x and y, the ratio λ becomes $1/\lambda$, since the meaning of δ and ε will also be interchanged.

We note that if λ becomes very large ($\lambda = \infty$), we are back to the classical case (often referred to as the "regression of y on x"); if, on the other hand, λ approaches zero, we obtain the same results as we would have obtained by interchanging x and y and using the classical formulas after this interchange. This is called the "regression of x on y."

Finally, we make the following important observation. One finds occasionally in the literature the recommendation that the regression of y on x (the use of the classical formulas) is justified even when x is subject to experimental error, provided that the "errors of x are small with respect to the errors of y." This is a meaningless statement, because the errors depend of course on the units used for x and y. If, for example, the original units of x are meters, and the errors of x are small with respect to those of y, a change of units of x from meters to centimeters will multiply all x errors by 100, and the latter may no longer be small with respect to the y-errors. The correct criterion is therefore not λ, but rather the ratio of the slope $\hat{\beta}$ to the square root of λ. This ratio is called the "sensitivity of y with respect to x". Denoting it by the symbol $\psi_{y/x}$, we have by definition

$$\psi_{y/x} = \frac{\beta}{\sqrt{\lambda}} = \frac{\beta}{\sigma_\varepsilon/\sigma_\delta} \tag{5.40}$$

It is easily seen that the value of ψ is unaffected by a change of units of either x or y, or both (see below). Furthermore, a *small* value of $\psi_{y/x}$ indicates that the errors of x, *in a true invariant scale,* are small with respect to those of y, and consequently the classical assumptions are satisfied. This can be explained as follows.

The scale in which x or y are expressed should logically have no effect on how the line is fitted; obviously a conversion from inches to feet, or from pounds to kilograms should not basically affect the fit. Hence, we can express x and y in such units that the slope of the line is unity. In such units, $\psi_{y/x}$ becomes equal to $\sigma_\delta/\sigma_\varepsilon$. Hence a small ψ-value indicates an error of x, (σ_δ), considerably smaller than the error of y, (σ_ε), in equivalent scales for x and y. Now, if we change the units of, say, x, so that in the new units, say x', we have $x' = kx$, then the slope in those new units becomes $\Delta y/\Delta x' = \Delta y/(k\ \Delta x) = (1/k)(\Delta y/\Delta x)$. The new ratio of standard deviations becomes $\sigma_y/\sigma_{x'} = \sigma_y/(k\sigma_x) = (1/k)(\sigma_y/\sigma_x)$. We see that the sensitivity in the new units, say ψ', becomes

$$\psi'_{y/x'} = \frac{\Delta y/\Delta x'}{\sigma_y/\sigma_{x'}} = \frac{\frac{1}{k}\beta}{\frac{1}{k}\ \sigma_y/\sigma_x} = \frac{\beta}{\sigma_y/\sigma_x} = \psi_{y/x} \tag{5.41}$$

so that the conclusion (small ψ entails small x-error in comparison to y-error) remains the same. Since $\psi_{y/x}$ is only *partly* determined by the ratio of the error variances (see Eq. (5.40)), the other determining factor being the slope β, λ by itself cannot be a valid criterion for judging the validity of the classical formulas when x is also subject to error. The proper criterion is the invariant quantity $\psi_{y/x}$. It can be verified that even a more complex change of units, such as degrees Celsius to degrees Fahrenheit, or even concentration of hydrogen ions to pH, leaves $\psi_{y/x}$ invariant.

5.6 PREDICTIONS WHEN BOTH VARIABLES ARE SUBJECT TO ERROR

The symmetry we have just mentioned allows us not only to use the regression equation as a valid calibration device, but also to estimate the precision of the estimate under realistic assumptions about the errors.

After running the calibration experiment, assume that we make a measurement on a "new" sample by the new method and obtain a value y_n. Then we obtain an estimated value of x, say $\hat{x}$, given by the equation

$$\hat{x} = \frac{y_n - \hat{\alpha}}{\hat{\beta}}$$

where $\hat{\beta}$ is given by Eq. (5.31) and $\hat{\alpha}$ by Eq. (5.32).

It is shown in Mandel (1984) that the standard deviation of $\hat{x}$ is given by the equation:

$$s_{\hat{x}} = \left[\frac{s_e^2}{\hat{\beta}^2} + \frac{D^2}{\hat{\beta}^2}\left(\frac{1}{N} + \frac{P^2\,(y_n - \bar{y})^2}{\hat{\beta}^2 Q}\right)\right]^{1/2}$$

To demonstrate the symmetry of our treatment with regard to x and y, we can interchange the meaning of these two symbols. Thus we can denote the well-established method by y and the new method by x. The value of λ would now be $1/\lambda$, or $1/576 = .001736$. If we measure a new sample, we may call the result x_n. Then the corresponding value by the well-established method $\hat{y}$ will be

$$\hat{y} = \hat{\alpha}^* + \hat{\beta}^* X_n$$

where α^* and β^* are obtained by formulas (5.31) and (5.32). All other equations are also the same as before, except the one for the estimation of the estimation of $S_{\hat{y}}$, which will be

$$s_{\hat{y}} = \left[\ \hat{\beta}^2\, s_d^2 + \left(\frac{1}{N} + \frac{P^2(x_n - \bar{x})^2}{Q}\right) D^2\right]$$

5.7 SUMMARY

We have reviewed three important cases of fitting a straight line: the "classical" case; the weighted fit, and the fit for the case in which both x and y are subject to error. We have emphasized the true motivation for weighting, which is not so much to get better estimates for the slope and intercept of the fitted line, as it is to obtain valid estimates of the error variances. Formulas have been given, and illustrative examples have been discussed.

REFERENCES

Brown, J. E., M. Tryon, and J. Mandel (1963). Determination of Propylene in Ethylene-Propylene Copolymers by Infrared Spectrometry. Anal. Chem., *35*, 2172–2176.

Hicho, G. (1983). Private communication, National Bureau of Standards, Washington, D.C.

6

Linear Functions of Several Variables

6.1 SCOPE

We have learned how to fit a linear relation to a set of points (x,y) under a variety of assumptions. Now we turn to the more general situation in which a measured value y is assumed to be a linear function of p quantities x_i ($i = 1$ to p). The computations, while well known, are laborious, and a computer is almost a prerequisite. The most interesting aspect of this problem is the interpretation of the outcome of the analysis. We deal in depth with this and other aspects of the fitting of linear functions to date.

6.2 THE GENERAL THEORY OF MULTIPLE LINEAR REGRESSION

We have already observed that the term *regression* is widely used in statistics to denote the fitting of equations to data.

Table 6.1 Schematic for Multiple Regression

Point	Regressors				Response
	x_1	x_2	. . .	x_p	y
1	x_{11}	x_{12}	. . .	x_{1p}	y_1
2	x_{21}	x_{22}	. . .	x_{2p}	y_2
3	x_{31}	x_{32}	. . .	x_{3p}	y_3
⋮			⋮		⋮
N	x_{N1}	x_{N2}	. . .	x_{Np}	y_N

We consider a relation of the type

$$y = \beta_1 x_1 + \beta_2 x_2 + \cdots + \beta_p x_p + \varepsilon \tag{6.1}$$

where $x_1, x_2, \ldots, x_p$ are the "independent" variables (for example, x_1 could be temperature, x_2 pressure, etc.) that can take on different values; y is a measurement made when $x_1, x_2, \ldots$ are given certain numerical values, and ε is the experimental error of the measured y-value. The variable y is called the *dependent* variable, or the *response*. The x variables are also called *regressors*. The β are unknown coefficients, to be estimated by the analysis.

While the adjective *independent* is often used in reference to the regressors, it is not to be taken in its common mathematical, or statistical sense. Regressors may, for example, include both x and x^2. The theory of regression treats them however as independent quantities, as we will soon learn. The data of a multiple linear regression appear in the form shown schematically in Table 6.1. In each row, the x-variables have been given specific numerical values (generally different from row to row), and y has been measured for this combination of x-values. The number of rows, N, must be at least equal to p, the number of regressors. Indeed, if $N < p$, there are fewer equations than unknowns. If $N > p$, there is in general a certain amount of inconsistency between the N equations, but this is attributed to the errors of measurement ε.

The object of the analysis is to resolve these inconsistencies, by finding a single set of β-values which, in some sense, are the "best" values possible.

In a great number of instances, the first regressor, x_1, has the value "one" for all N equations. The reason for this is to allow for an "independent term" in the model equation (6.1). For example, in the case $p = 2$, if $x_1 \equiv 1$, we have

$$y = \beta_1 + \beta_2 x_2 + \varepsilon$$

which is of course exactly the straight line fitting situation we have discussed in Section 5.2, now expressed in slightly different symbols. We see that Eq. (6.1) is in some sense a generalization of the straight line fit.

The solution of our problem is again accomplished by applying the least squares principle.

Suppose that the solution consists of a set of numerical values (estimates).

$$\hat{\beta}_1, \hat{\beta}_2, \ldots, \hat{\beta}_p$$

Then, the estimated value of y_i, denoted $\hat{y}_i$, for the set of the p x-values occurring in the ith row of the table is given by

$$\hat{y}_i = \hat{\beta}_1 x_{i1} + \hat{\beta}_2 x_{i1} + \cdots + \hat{\beta}_p x_{ip} \tag{6.2}$$

This value will in general be different from the observed value y_i. Again we define the "residual" by

$$d_i = y_i - \hat{y}_i \tag{6.3}$$

There will be one residual for each row of the table. The principle of least squares requires that the sum of squares of these N residuals be minimum:

$$\Sigma_i \, d_i^2 = \text{minimum} \tag{6.4}$$

Equation (6.4) can be written as:

$$\Sigma_i \, (y_i - \hat{\beta}_1 x_{i1} - \hat{\beta}_2 x_{i2} - \cdots - \hat{\beta}_p x_{ip})^2 = \text{minimum} \tag{6.5}$$

The summation is over all N rows of the table. Calculus teaches us that the minimum is obtained by taking derivatives with respect to each unknown (each β) and setting these derivatives equal to zero. This gives p equations (one for each unknown), of the form

$$2\Sigma_i \, (y_i - \hat{\beta}_2 x_{i1} - \hat{\beta}_2 x_{i2} - \cdots - \hat{\beta}_k x_{ik} - \cdots - \hat{\beta}_p x_{ip}) x_{ik} = 0 \tag{6.6}$$

or, dividing by 2, and writing d_i for the quantity in parentheses (because this is the ith residual, by definition):

$$\Sigma\ d_i x_{ik} = 0 \tag{6.7}$$

where the summation is over all N rows. It is instructive to write Eq. (6.6) in the form

$$\Sigma\ y_i x_{ki} = \hat{\beta}_1\ \Sigma_i\ x_{i1} x_{ik} + \hat{\beta}_2\ \Sigma_i\ x_{i2} x_{ik} + \cdots + \hat{\beta}_p\ \Sigma_i\ x_{ip} x_{ik} \tag{6.8}$$

There will be p such equations, one for each k-value. The construction of these p equations, called the *normal equations* is seen to proceed in accordance with the following rule:

Select, in succession, each of the p columns in the table of "observational" equations (Table 6.1). For each selected column, say column k, multiply each element of the ith row by x_{ki}, and sum over all rows. Performing this operation for each k value we obtain the p normal equations in the p unknowns $\hat{\beta}_1, \ldots, \hat{\beta}_p$. They are then solved in the usual way by algebraic methods.

We will not discuss these computational matters, which can be found in books on numerical analysis, and are, nowadays, available in many commercially available computer software packages. The numerical effort involved in solving the normal equations (say for $p \geqslant 3$) is such that a computer is almost indispensable, except in some very special situations. However, several interesting features of multiple linear regression will be discussed further in this chapter.

6.3 ORTHOGONALITY

Equation (6.8) involves, when expanded, sums of the form $\Sigma\ x_j x_k$, where the sum extends over all N rows. This can be written more explicitly as:

$$\Sigma_i\ x_{ij} x_{ik} \tag{6.9}$$

where i is the row index ($i = 1$ to N) and i and k denote two specific columns. For each k, one of the js will be k itself, in which case Eq. (6.9) becomes

$$\Sigma_i \, x_{ik}^2 \tag{6.10}$$

Now suppose that for one of the x variables, say x_k, we have

$$\Sigma_i \, x_{ij} x_{ik} = 0 \qquad \text{for all } j, \text{ except } j = k \tag{6.11}$$

(The exception is necessary since $\Sigma_i \, x_{ik}^2$ cannot be zero unless each $x_{ik} = 0$ for all i, which is tantamount to saying that the variable x_k is absent from the regression.)

If Eq. (6.11) holds, then x_k is said to be "orthogonal" to all other x_j. But if this is the case, then Eq. (6.8), for that particular value of k, will reduce to

$$\Sigma_i \, (y_i x_{ik} - \hat{\beta}_k x_{ik}^2) = 0$$

or

$$\hat{\beta}_k = \frac{\Sigma_i \, y_i x_{ik}}{\Sigma_i \, x_{ik}^2} \tag{6.12}$$

Thus, if x_k is orthogonal to all other x_j, then $\hat{\beta}_k$ can be calculated knowing x_k only, without involvement of any of the other x_j variables. We will make use of Eqs. (6.7) and (6.12) in the following section.

6.4 FITTING A QUADRATIC EQUATION BY ORTHOGONAL POLYNOMIALS

Table 6.2 presents measurements of the density of aqueous solutions of methanol at six different concentrations, at a fixed temperature of 40°C. If we fit a straight line to this set of data, we obtain the residuals shown in the table. They clearly indicate curvature. Consequently a straight line fit is unsatisfactory. A plausible alternative is to try a "quadratic" fit, given by the relation

$$y = \alpha + \beta c + \gamma c^2 + \varepsilon \tag{6.13a}$$

where y denotes density and c, concentration. In general, the coefficients α, β, and γ would be calculated by the numerical analysis

Table 6.2 Density of Aqueous Solutions of Methanol

Data concentration, % (c)	Density (y)	Residual (d)	Orthogonal polynomials P_0	P_1	P_2
4.91	0.9833	−0.0084	1	−46.381	1204.60
22.92	0.9536	0.0021	1	−28.371	−185.05
39.99	0.9199	0.0065	1	−11.301	−903.35
59.98	0.8742	0.0054	1	8.689	−1003.70
80.04	0.8257	0.0017	1	28.749	−301.00
99.91	0.7723	−0.0073	1	48.619	1188.46

$\hat{\gamma}_0 = .88817$, $\hat{\gamma}_1 = 0.002233$, $\hat{\gamma}_3 = -0.000006463$.

techniques mentioned above. In doing this, c and c^2 are considered separate, "independent" variables; they could be called x_1 and x_2 respectively, without in any way affecting the fitting procedure. We will however present a different technique, known as the *method of orthogonal polynomials*.

The numerical results will be the same, and the method we are about to present is *not* recommended for persons having the use of a computer and of an appropriate software package at their disposal. The method is, however, intrinsically interesting and touches upon some important facts in regression analysis. Furthermore, it can be carried out on any hand calculator that is programmed for straight line fitting.

The method of orthogonal polynomials applies to the fitting of *polynomials* of the type

$$y = \alpha + \beta x + \gamma x^2 + \delta x^3 + \cdots + \varepsilon$$

We illustrate it for fitting the quadratic given by Eq. (6.13).

As the name suggests, the variables $x_1 = 1$, $x_2 = c$, $x_3 = c^2$ are replaced by polynomials $P_0(c)$, $P_1(c)$, $P_2(c)$ which satisfy the following two conditions:

$$\text{(a)} \begin{cases} P_0(c) \text{ is of degree 0; i.e., it is a constant.} \\ P_1(c) \text{ is of degree 1.} \\ P_2(c) \text{ is of degree 2.} \end{cases}$$

The quadratic equation can now be written in the form

$$y = \gamma_0 P_0(c) + \gamma_1 P_1(c) + \gamma_2 P_2(c) + \varepsilon \tag{6.13b}$$

(b) All polynomials $P_k(c)$ are mutually orthogonal. Without proof, we state that

$$P_0(c) = 1 \qquad P_1(c) = c - \bar{c} \tag{6.14}$$

We verify that P_0 and P_1 are orthogonal: indeed,

$$\Sigma\, P_0(c_i)P_1(c_i) = \Sigma_i\, 1 \cdot (c_i - \bar{c}) = 0$$

Therefore, Eq. (6.12) applies and we have

$$\hat{\gamma}_0 = \frac{\Sigma\, yP_0(c)}{\Sigma\, (P_0(c))^2} = \frac{\Sigma\, y}{N} = \bar{y} \tag{6.15}$$

$$\hat{\gamma}_1 = \frac{\Sigma\, yP_1(c)}{\Sigma\, (P_1(c))^2} = \frac{\Sigma\, y(c - \bar{c})}{\Sigma\, (c - \bar{c})^2} \tag{6.16}$$

We will now attempt to find $P_2(c)$. The model equation can be written

$$y = \gamma_0 P_0(c) + \gamma_1 P_1(c) + \gamma_2 P_2(c) + \varepsilon \tag{6.17}$$

To simplify the notation, let us represent $c_i - \bar{c}$ by z_i:

$$z_i = c_i - \bar{c}$$

and let

$$P_2(c) = z_i^2 + az_i + b \tag{6.18}$$

Orthogonality of P_2 with P_0 requires $\Sigma\, z_i^2 + a\, \Sigma\, z_i + N \cdot b = 0$. But $\Sigma\, z_i = 0$; hence

$$b = -\frac{\Sigma\, z_i^2}{N} = -\frac{S_2}{N} \tag{6.19}$$

where $S_2 = \Sigma\, z_i^2 = \Sigma\, (c_i - \bar{c})^2$.

Orthogonality of P_2 with P_1 requires $\Sigma\, z_i^3 + a\, \Sigma\, z_i^2 + b\, \Sigma\, z_i = 0$, hence

$$a = -\frac{\Sigma\, z_i^3}{\Sigma\, z_i^2} = -\frac{S_3}{S_2} \tag{6.20}$$

where $S_3 = \Sigma\, z_i^3 = \Sigma\, (c_i - \bar{c})^3$.

Introducing Eqs. (6.19) and (6.20) into (6.18), we get

$$P_2(c) = z_i^2 - z\frac{S_3}{S_2} - \frac{S_2}{N} \tag{6.21}$$

Since $z_i = c_i - \bar{c}$, we can readily calculate all three P polynomials for each i:

$$P_0 = 1$$

$$P_1 = c_i - \bar{c}$$

$$P_2 = (c_i - \bar{c})^2 - (c_i - \bar{c})\frac{S_3}{S_2} - \frac{S_2}{N}$$

Now, because of the orthogonality of the Ps, Eq. (6.12) applies and we get

$$\hat{\gamma}_0 = \frac{\Sigma\, y_i P_{i0}}{\Sigma\, P_{i0}^2} = \bar{y}$$

$$\hat{\gamma}_1 = \frac{\Sigma\, y_i P_{i1}}{\Sigma\, P_{i1}^2}$$

$$\hat{\gamma}_2 = \frac{\Sigma\, y_i P_{i2}}{\Sigma\, P_{i2}^2}$$

The calculations are rather simple. The orthogonal polynomials for our example are also shown in Table 6.2, as well as the coefficients γ_0, γ_1, and γ_2. We have already mentioned that this method of calculation is in general not the most efficient. On the other hand, it has definite advantages. Among those is the possibility to obtain from the γ-coefficients, and the orthogonial polynomials, both the linear fit and the quadratic fit. This is shown in Table 6.3 which lists the fitted y-values as well as the residuals for both fits, and shows, by the formulas at the bottom, how they were obtained. We see, incidentally, that the quadratic fit is, for these data, far superior to the linear one, because of the presence of curvature.

A second advantage of the method of orthogonal polynomials is that in the case where several responses y have to be fitted to the same set of x values, the calculations are vastly simplified.

Table 6.3 Linear and Quadratic Fits for Data of Table 6.2

Linear fit		Quadratic fit	
$\hat{y}$[a]	$10^4 \cdot d$	$\hat{y}$[b]	$10^4 \cdot d$
0.99173	−84.3	0.98395	−6.5
0.95152	20.8	0.95271	8.9
0.91340	65.0	0.91923	6.7
0.86877	54.3	0.87524	−10.4
0.82398	17.2	0.82591	−2.1
0.77961	−73.1	0.77191	3.7

[a]Linear fit: $\hat{y} = \hat{\gamma}_0 \cdot P_0 + \hat{\gamma}_1 \cdot P_1$.
[b]Quadratic fit: $\hat{y} = \hat{\gamma}_0 \cdot P_0 \cdot \hat{\gamma}_1 + P_1 + \hat{\gamma}_2 \cdot P_2$.

6.5 COLLINEARITY

The example in the previous section is an illustration of the treatment of mathematically related quantities, such as c and c^2, as "independent," in the sense that the *computational* part of linear regression treats them as though they were unrelated. The *interpretation* of the regression equation, however, unlike the computational aspects, is very much affected by the lack of independence. To see this, we now look upon a regression situation in the following geometrical way.

To fix the ideas, consider a regression with *two* regressors x_1 and x_2. Thus the equation is

$$y = \beta_1 x_1 + \beta_2 x_2 + \varepsilon \tag{6.22}$$

Geometrically, we can represent it in a three-dimensional space, in which x_1 and x_2 are plotted along two perpendicular axes, drawn, for example, on this sheet of paper. The response y is plotted along an axis perpendicular to this plane and going through the origin of the (x_1, x_2) system. Each combination of x_1 and x_2 is then a point in the base-plane, and y is the length of a straight line segment perpendicular to the plane at the point (x_1, x_2). The end points of all y-segments, for all (x_1, x_2) points, corrected for the error ε, will then lie in a plane. This is merely a consequence of the linear nature of Eq. (6.22). Therefore, fitting Eq. (6.22) to a set of (x_1, x_2) points is really fitting a plane surface in three dimensional space.

In general, the points (x_1, x_2) are selected according to the experimental plan underlying the experiment. For example, they may be the four corners of a rectangle, such as (1,1), (1,5), (8,1), and (8,5). If the equation is, for example,

$$y = 2x_1 + 3x_2 + \varepsilon$$

then, the vertical segments representing y will be erected at the four points mentioned above, and have, apart from experimental error, lengths of respectively, 5, 17, 19, and 31.

Now consider the situation we discussed in Section 6.3. There we had, apart from the constant term, also two regressors, but they were c and c^2 respectively. Calling them respectively x_1 and x_2, the points (x_1, x_2) now *cannot* be selected according to the experimenter's desire; once we have selected a value for x_1 (c), x_2 is thereby automatically determined ($x_2 = c^2 = x_1^2$). Thus, the *design points,* as they are often called, lie in this case *along a curve* expressed by $x_2 = x_1^2$. All vertical segments y have their foot on this curve. The surface fitted is still a plane, but the only part of that plane that is relevant is the curve that forms the intersection of this plane with the set of vertical y-segments whose feet are on the (x_1, x_1^2) curve. By the very nature of the problem, we would not think of calculating a value of y for, for example, $x_1 = 2$ and $x_2 = 10$, since for $x_1 = 2$, x_2 must be 4.

We can express this situation in the following way. Suppose that the range of interest is $x_1 = 2$ to 5. Then x_2, which is equal to x_1^2, covers the range 4 to 25. All points fall in a rectangle in the x_1, x_2 plane whose corners (2,4), (2,25), (5,4), (5,25). We may call this the *sample domain*. Most of this sample domain is however empty, since only the points on the curve $x_2 = x_1^2$ are of interest. If after the fitting process we wish to "predict" the value of y for any point, it will have to be a point on this curve. We can refer to the totality of these points as the *effective prediction domain,* or EPD. The EPD in this sample is only a small portion of the sample domain, namely the curve $x_2 = x_1^2$ between the points (2,4) and (5,25).

There are cases where x_1 and x_2 are *linearly* (rather than nonlinearly) related. For example, we may have $x_2 = 5 + 4x_1$.

In that case, the EPD will consist of a *straight line segment* inside the sample domain. But the most interesting case arises, in real life situations, where x_1 and x_2 are only *approximately* linearly related. As an example, suppose that a study is made of the grades obtained by school children of a certain age, as a function of the education of their parents and of their financial status. Thus: x_1 = education of parents; x_2 = financial status; y = grade of child.

Now, while the financial status of a family is not mathematically related to the education of the head of the family and his (or her) spouse, it is to be expected that a positive correlation exists between these variables. But then the (x_1, x_2) points (the design points) will tend to fall inside an elongated ellipse in the sample domain, stretching from its lower left corner to its upper right corner. The question of interest is now as follows: after fitting the regression equation, that is, finding the $\hat{\beta}$ in the equation

$$\hat{y} = \hat{\beta}_0 + \hat{\beta}_1 x_1 + \hat{\beta}_2 x_2$$

what use can legitimately be made of this "prediction equation"? Can we, for example, use it to predict the grade of a child whose parents are very affluent but very poorly educated? Such a family was presumably not present in the sample covered by the study, and presents therefore a new case for which no reliable data are available. A sound scientific approach would dictate that such inferences not be made, because in science one does not "predict" results for unstudied phenomena. Of course, one often makes such predictions tentatively, *as hypotheses* to be *tested* by further experimentation, but not as final conclusions.

We take the position here that predictions should be confined to the area actually covered by the experiment, that is, to the EPD as defined above. In the case just described, the EPD would be the ellipse within which the design points fall, or an area that is approximately equivalent to it. This would leave substantial "holes" in the sample domain, but it would be the proper scientific inference.

The word *collinearity,* or *multicollinearity,* has been used to describe situations of the type just discussed. *Exact* collinearity exists when two or more of the regressors are linearly related (as, for example, in $x_2 = 5 + 4x_1$) by an exact linear relation. Approximate collinearity or, simply "collinearity," exists when a *correlation* (rather than an exact linear relation) exists between two or more regressors (as in the example of the grades of school children).

Attempts have been made to "remedy" collinear data by various mathematical devices. We believe that this is unadvisable and that the correct treatment of collinearity consists in establishing the EPD and then stating that valid predictions can be made only for points falling in the EPD or, at least, not too far from it.

The calculation of the EPD in the general case is beyond the scope of this book. We can however cover the simple case of two regressors x_2 and x_3, in addition to the regressor $x_1 = 1$:

$$y = \beta_1 \cdot 1 + \beta_2 x_2 + \beta_3 x_3 + \varepsilon \tag{6.23}$$

We will illustrate the analysis by means of an example. The notation in Eq. (6.23) is consistent with that of Eq. (6.1), in which $x_1 = 1$ for all points.

6.6 TREATMENT OF COLLINEARITY FOR A PARTICULAR EXAMPLE

The example we choose is a portion of a data set discussed by Fearn (1983) and dealing with the calibration of a near infrared reflectance instrument for the measurement of protein in ground wheat samples. Fearn reports the results on 24 samples of ground wheat for each of which protein content was measured in two ways: by a chemical method based on the determination of nitrogen, and by a physical method based on the determination of reflectance at six different wavelengths. The chemical method is the reference method; it is quite laborious. The underlying idea is that the chemical result can be expressed as a linear combination of the six physical measurements, so that given the set of six reflectances for any new, "unknown" sample of ground wheat, one can predict the result that would have been obtained by the chemical method.

For simplicity of presentation, we have chosen a portion of the complete data, consisting of only 12 samples, and limited to the measurement at only two wavelengths. This reduced data set is shown in

Table 6.4 Protein in Ground Wheat (Partial Data)

Data				Results of regression[a]	
x_1	x_2	x_3	y	$\hat{y}$	d
1	246	374	9.23	9.10	0.13
1	269	389	11.35	11.53	−0.18
1	240	359	10.95	10.94	0.01
1	236	352	11.67	11.50	0.17
1	243	366	10.41	10.13	0.28
1	273	404	9.51	9.18	0.33
1	242	370	8.67	8.99	−0.32
1	238	370	7.75	7.98	−0.23
1	258	393	8.05	7.87	0.18
1	264	384	11.39	11.39	0.00
1	293	421	10.23	10.41	−0.18
1	242	366	9.70	9.88	−0.18

[a] $\hat{\beta}_1 = 30.8532$, $\hat{\beta}_2 = 0.2512$, $\hat{\beta}_3 = -0.2234$. Standard deviation of fit $= s_f = 0.239$.

the first four columns of Table 6.4. The model is that of Eq. (6.23) in which x_2 and x_3 are the reflectance measurements at the two wavelengths, and y is the chemical measurement.

6.6.1 Results of the Multiple Regression Calculations

We first report the results obtained by the application of the general theory outlined in Section 6.1. They are shown in Table 6.4, and consist of the estimates of β_1, β_2, and β_3, as well as the estimates $\hat{y}$, the residuals, and the standard deviation of fit. In spite of first appearances, this is by no means "the end of the story." The question that now arises is: what are the limitations, if any, to the use of the regression equation

$$y = 30.8532 + .2512x_2 - .2234x_3 \tag{6.24}$$

for "future" samples of ground wheat?

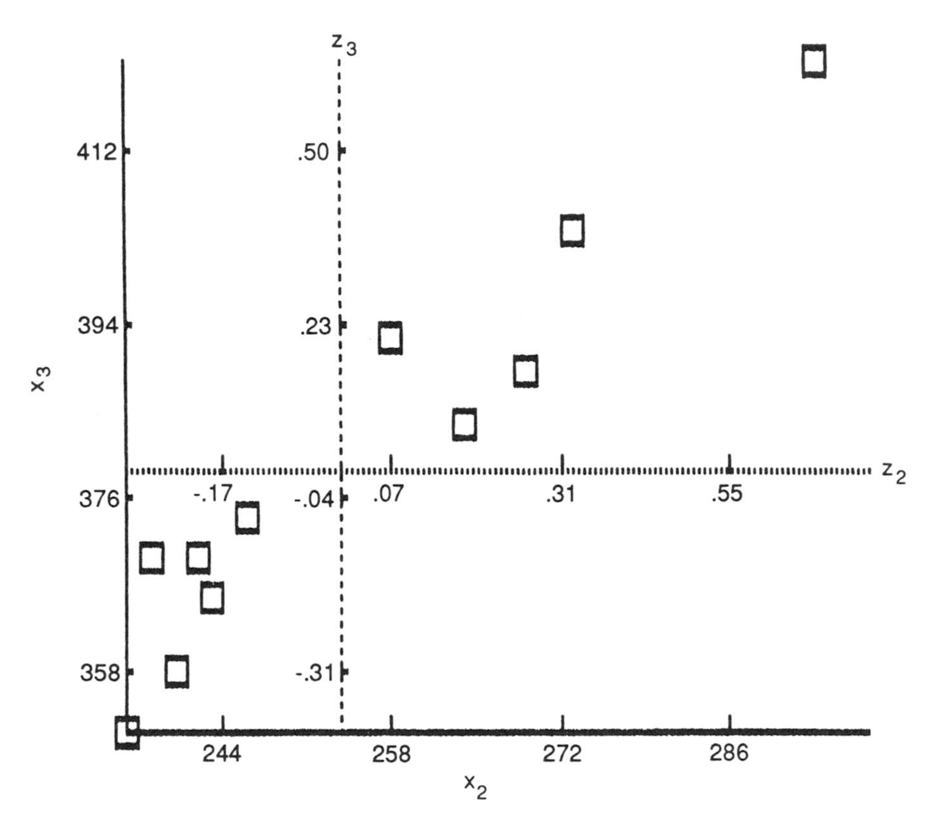

Figure 6.1 Protein in ground wheat (Table 6.4), x_3 versus x_2.

6.6.2 Collinearity and EPD

To answer this question, we have to construct an EPD of the x-variables. We proceed in five steps:

Transformation to the Z-scale

Figure 6.1 is a graph of x_3 versus x_2 for the 12 samples of wheat. Our first step consists in moving the coordinate axes, parallel to themselves, to a point located in the center of the cluster of (x_2, x_3) points, and at the same time, changing the scales on x_2 and x_3 to a "standard" scale.

Table 6.5 z and w Coordinates for Data of Table 6.4

z_1[a]	z_2[a]	z_3[a]	w_a	w_b	w_c
0.2887	−0.1301	−0.0757	0.2887	−0.1455	0.0385
0.2887	0.2603	0.1514	0.2887	0.2911	−0.0770
0.2887	−0.2320	−0.3028	0.2887	−0.3781	−0.0500
0.2887	−0.2999	−0.4087	0.2887	−0.5011	−0.0770
0.2887	−0.1811	−0.1968	0.2887	−0.2672	−0.0111
0.2887	0.3282	0.3784	0.2887	0.4996	0.0355
0.2887	−0.1980	−0.1362	0.2887	−0.2364	0.0437
0.2887	0.2659	−0.1362	0.2887	−0.2844	0.0917
0.2887	0.0736	0.2119	0.2887	0.2019	0.0978
0.2887	0.1754	0.0757	0.2887	0.1775	−0.0705
0.2887	0.6677	0.6358	0.2887	0.9217	−0.0225
0.2887	−0.1980	−0.1968	0.2887	−0.2792	0.0009
			$\lambda_a = 1.0000$	$\lambda_b = 1.9574$	$\lambda_c = 0.0426$

$\lambda_1 = 1.9574$, $\lambda_2 = 1.000$, $\lambda_3 = 0.0426$. $w_1 = w_b$, $w_2 = w_a$, $w_3 = w_c$.
[a]$Z_1 = 1/\sqrt{12}$, $z_2 = (x_2 - 253.6667)/58.9124$, $z_3 = (x_3 - 379)/66.0606$ (see Eq. (6.25)).

Both of these objectives are accomplished by calculating for each of the 24 "points" (samples of wheat), the quantities

$$z_2 = \frac{x_2 - \bar{x}_2}{\sqrt{\Sigma\,(x_2 - \bar{x}_2)^2}} \qquad z_3 = \frac{x_3 - \bar{x}_3}{\sqrt{\Sigma\,(x_3 - \bar{x}_3)^2}} \tag{6.25}$$

The z-axes are also shown in Fig. 6.1, and are seen to pass through a new "origin" consisting of the point $(\bar{x}_2, \bar{x}_3)$. Furthermore, the scale on each z-axis is such that the two variables are now both expressed in units that are proportional to their respective standard deviations. For our data the latter property is not a striking feature, since the two standard deviations were already very similar in the x-scale but in many situations the different regressors are properties measured in entirely different scales. In such cases the z-transformation to comparable scales is very useful. In addition to the new variables z_2 and z_3, we consider also a new variable z_1, to correspond to $x_1 = 1$. This variable, z_1, is simply equal to $1/\sqrt{N}$, where N is the number of points ($N = 12$ for our data). The results of these calculations are shown in the first three columns of Table 6.5.

Rotation of the Axes

The transformation to the z-scales has not altered the basic nature of the regressors and their relation to each other. Figure 6.1 shows a definite tendency to collinearity. Our next step consists in a rotation of the system of coordinate axes, in such a way that the first axis will tend to follow the pattern of the (x_2, x_3) points, and the second axis will be perpendicular to that trend. The new coordinate system is denoted by the letter w, so that we now have, including w_a, which we define to be identical with z_1, three variables w_a, w_b, w_c. The reason for the subscripts a, b, and c will soon become apparent. The calculation of the new coordinates is quite simple, and is given by the following equations:

$$w_a = 1/\sqrt{N} \tag{6.26a}$$

$$w_b = (z_2 + z_3)/\sqrt{2} \tag{6.26b}$$

$$w_c = (z_3 - z_2)/\sqrt{2} \tag{6.26c}$$

The values of the z- and w-coordinates are shown in Table 6.5, for all 12 points.

The Eigenvalues and the w-Ellipsoid

We will not define the term *eigenvalue,* sometimes also called *latent root,* but indicate what it means in the present context (for more detail, see Mandel, 1982).

At the bottom of each w-column, a number, denoted λ, is given. It is simply the sum of squares of all w-values in that column. These three λ-values are the eigenvalues for the rotation to the w-coordinate system. They have a simple geometric interpretation. As we mentioned earlier, the totality of the (x_1, x_2, x_3) points of the experiment can be considered to lie in an ellipsoid of three dimensions. The λ are the squares of the half-lengths of the ellipsoid-axes (analogous to the square of the radius of a circle). They are, in fact, the parameters in the equation of the ellipsoid

$$\frac{w_a^2}{\lambda_a} + \frac{w_b^2}{\lambda_b} + \frac{w_c^2}{\lambda_c} = 1 \tag{6.27}$$

This equation holds for points on the surface of the ellipsoid. For points *inside* the ellipsoid we have

$$\frac{w_a^2}{\lambda_a} + \frac{w_b^2}{\lambda_b} + \frac{w_c^2}{\lambda_c} < 1$$

and for points *outside* the ellipsoid we have

$$\frac{w_a^2}{\lambda_a} + \frac{w_b^2}{\lambda_b} + \frac{w_c^2}{\lambda_c} > 1$$

The reason for the subscripts a, b, c is that it is customary and useful to consider the w-coordinates in the order of *decreasing* λ-value. It is only *after* calculating the w and λ that we can establish this order. In our example, we find (see Table 6.5) that the order is $\lambda_b > \lambda_a > \lambda_c$. Therefore, we state

$$w_1 = w_b \qquad w_2 = w_a \qquad w_3 = w_c$$

$$\lambda_1 = \lambda_b \qquad \lambda_2 = \lambda_a \qquad \lambda_3 = \lambda_c$$

We call the expression on the left side of Eq. (6.27) the *variance factor,* and denote it by VF:

$$\text{VF} = \frac{w_1^2}{\lambda_1} + \frac{w_2^2}{\lambda_2} + \frac{w_3^2}{\lambda_3} \tag{6.28}$$

The reason for this terminology will be explained shortly.

Interpretation of the w-Ellipsoid

One way of defining an EPD is to equate it with the ellipsoid. Indeed, since λ is the sum of squares of the w values corresponding to it, it is evident that the w^2 for each observed point is less than (or at most equal to) the corresponding λ. Thus, for the w-value of each observed point, for each axis, we have

$$-\sqrt{\lambda} \leqslant w \leqslant +\sqrt{\lambda} \tag{6.29}$$

which means that each point *of the experiment* lies inside (or, at most, on the surface of) the ellipsoid. It is also readily understood that the ellipsoid is, in some sense, the most compact area that includes *all* points of the experiment. Let us now observe that, for our data, the

Table 6.6 Calculation of Variance Factor for Corners of Sample Domain (Data of Table 6.4)

Point	x_1	x_2	x_3	w_1	w_2	w_3
1	1	236	352	−0.5011	0.2887	−0.0770
2	1	236	421	0.2375	0.2887	0.6616
3	1	293	352	0.1831	0.2887	−0.7611
4	1	293	421	0.9217	0.2887	0.0225

Point	$\frac{w_1^2}{\lambda_1}$	$\frac{w_2^2}{\lambda_2}$	$\frac{w_3^2}{\lambda_3}$	VF
1	0.1282	0.0833	0.1390	0.3506
2	0.0288	0.0833	10.2754	10.3875
3	0.0171	0.0833	13.5783	13.6988
4	0.4339	0.0833	0.0119	0.5292

λ_3 value is an order of magnitude smaller than λ_1 and λ_2. Thus, the ellipsoid is extremely narrow in the direction of the λ_3-axis. Consequently, it is in that direction that the EPD will be most likely to eliminate points for the purpose of useful prediction.

To further clarify this point, we consider the *corners* of the sample domain. These are the combinations of the smallest and largest values for each regressor other than the first (since $x_1 = 1$). There are four such points, given by the coordinates listed in Table 6.6. In Table 6.6 we also list, in addition to the w_j, the quantities w_j^2/λ_j for all j, together with their sum, the variance factor. The calculations leading to these values are straightforward: given a point (x_1, x_2, x_3) we convert it first to z-values, by means of Eq. (6.25) (remembering that $z_1 = 1/\sqrt{N}$) and then to w-coordinates by means of Eqs. (6.26).

Table 6.6 reveals two interesting facts. First, we see that two of the four corners lie well outside the EPD. This will also apply to all points in the vicinity of these two corners, leaving a considerable region of the sample domain unsuitable for prediction. Second, we note that in both cases, it is w_3 that causes the VF. In practice it is sufficient to concentrate on the w-values corresponding to very small λ-values, since they alone will cause the VF to be large.

Interpretation of the Variance Factor

Once the coefficients $\hat{\beta}$ have been calculated, a fitted value, $\hat{y}$, can be calculated for any set of x-values, and an estimate s_f, the standard deviation of fit, can be calculated. The square of this quantity, s_f^2, is calculated by the following formula:

$$s_f^2 = \frac{\Sigma d^2}{N - p} \tag{6.30}$$

where d is a residual and p the number of $\hat{\beta}$ values calculated. In our illustrative example, we have $N = 12$ and $p = 3$. The variance s_f^2 is a sample estimate of the variance of experimental error σ_ε^2.

Now, it can be shown that the variance of an estimated value $\hat{y}$ is equal to

$$\text{Var}(\hat{y}) = (\text{VF})_{\hat{y}} \cdot \sigma_\varepsilon^2 \tag{6.31}$$

where $(\text{VF})_{\hat{y}}$ is the variance factor at the point for which $\hat{y}$ was calculated. An estimate for $\text{Var}(\hat{y})$, which we denote by $s_{\hat{y}}^2$ is given by

$$s_{\hat{y}}^2 = (\text{VF})_{\hat{y}} \cdot s_f^2 \tag{6.32}$$

Now we know that the larger the VF, the further outside is the point with respect to the EPD. We now see that the result of being far outside is a large variance of $\hat{y}$. Thus, the further away a point is from the EPD, the larger is the uncertainty of the estimated $\hat{y}$ at this point. (Note that VF depends only on the x-values, not on y.)

This is one reason for refraining from using the regression equation as a predictor for points that are considerably outside the EPD. We may call it the "statistical" reason. An even more cogent reason, of a nonstatistical nature, is the following. The success of the fitting process is measured by the smallness of s_{fit}^2. If s_{fit}^2 is satisfactorily small, the regression equation fits the data well. It will then, in most cases, also be a good fit for all points inside the EPD. However, outside the EPD, the linear equation (Eq. (6.1)) that was postulated as a model, may no longer be valid. This is especially so if our fit is purely empirical, without a sound theoretical basis. Thus, extrapolation of the equation well outside the EPD is unjustified, first because there is no assurance that the model is still valid for the point considered and secondly because, even if the model were valid, the uncertainty of $\hat{y}$ may be too large for any practical use.

We now perceive the usefulness of calculating an EPD. It tells us when we are safe in using the regression equation and when such a use is risky. The VF is a useful measure for this purpose. For any proposed "point," that is, for any set (x_1, x_2, x_3), we can calculate the corresponding VF. If it is appreciably larger than unity, the use of the regression equation is risky. If VF ≤ 1, or not much greater than 1, the regression equation can be safely used for prediction. As an example, consider one of the points included in Fearn's data, but left out of Table 6.4. The x-coordinates of the point are (1, 276, 396). Its VF value is obtained by first calculating (z_1, z_2, z_3) and then (w_1, w_2, w_3); and finally, using the eigenvalues calculated from our data, the expression $w_1^2/\lambda_1 + w_2^2/\lambda_2 + w_3^2/\lambda_3$. We find, VF = 0.3608 and conclude that the use of the regression equation for this point is entirely acceptable.

The calculations involved may seem tedious. Indeed in any multiple regression work, the computer is almost indispensable, unless one is willing to spend many hours in tedious calculations. With the computer, on the other hand, the results can be obtained almost instantaneously.

An important point to be made is that collinearity is *not,* as is sometimes implied in the literature, a sort of "malady" of the data, to be remedied or "healed" by special mathematical devices (for example, "ridge regression"). When present, collinearity merely tells us that use of the regression equation may be risky, even for some areas *inside* the sample domain. The criterion to be used is to find where the point of proposed application lies with respect to the EPD. The VF is a useful device for this purpose (see Mandel, 1985).

6.7 SUMMARY

In this chapter we have dealt with both computational and interpretational aspects of fitting linear equations to data. We have covered the technique of orthogonal polynomials and have dealt in detail with the ubiquitous problem of collinearity.

REFERENCES

Fearn, T. (1983). A Misuse of Ridge Regression in the Calibration of a Near-Infrared Reflectance Instrument. Appl. Stat., *32,* 73–79.

Mandel, J. (1982). Use of the Singular Value Decomposition in Regression Analysis. Am. Stat., *36,* 15–24.

Mandel, J. (1985). The Regression Analysis of Collinear Data. J. Res. National Bureau of Standards, *90,* 465–476.

7

Structured Two-Way Tables of Measurements

7.1 FACTORIAL DESIGNS

We have discussed, in Chapter 6, the case of measurements y that depend on a number of quantities, called regressors, in terms of a relation of the type

$$y = \beta_1 x_1 + \beta_2 x_2 + \cdots + \varepsilon \tag{7.1}$$

where ε is the experimental error in the measurement of y.

In this chapter we deal with a situation which, in one way, is more restrictive, but, in another aspect, far more general. The restriction is that we consider only two independent regressors x_1 and x_2, for example, pressure and wavelength. The greater generality of our present case is that we are dealing with a *crossed* classification of x_1 and

x_2. By that we mean that for each value of x_1, we choose a set of values of x_2, *the same for all values of* x_1, and make a measurement of y for each resulting combination of x_1 and x_2. Table 7.1 presents an example of such a situation. A design of this type is called a complete *factorial* design, the reason being that y is measured at all combinations of the two *factors* x_1 and x_2. It is of course possible to construct factorial designs involving more than two factors, but we will not discuss them in this book.

Table 7.1 deals with measurements of the compressive strength of timber for ten trees (factor x_1) and five levels of temperature (factor x_2) (Williams, 1959). Each entry is the average of two measurements. The table also lists the row and column averages R_i and C_j and the grand mean, M. Instead of postulating a linear relation of the type of Eq. (7.1), we write the far more general model:

$$y_{ij} = F(x_1, x_2) + \varepsilon \tag{7.2}$$

In many cases, the function F is unknown before the experiment. Then, the purpose of the experiment is to provide us with a plausible function F. This is referred to as "fitting an empirical function to the data."

In other cases, we may have either partial or complete information

Table 7.1 Compressive Strength of Timber

Tree/ Temperature (°C)	−20	0	20	40	60	R_i
1	6570	6230	4715	3815	3170	4900
2	7950	7055	5650	4780	3635	5814
3	6695	6160	4825	3950	3205	4967
4	7755	6840	5165	4135	3530	5485
5	7765	6580	5145	4335	3340	5433
6	7630	6820	5175	4335	3310	5454
7	7530	6625	5280	4050	3075	5312
8	7605	6770	5230	4150	3045	5360
9	8450	7615	5970	4670	3130	5967
10	7725	7030	5370	3875	3145	5429
C_j	7568	6773	5253	4210	3259	$M = 5412$

on F, prior to the experiment. Then, the purpose of the experiment is either to provide the missing information or to confirm the prior knowledge about F.

7.2 TWO-WAY ANALYSIS OF VARIANCE

It is customary to analyze tables like Table 7.1 by a standard technique, called "two-way analysis of variance." In spite of its popularity, we do not recommend this technique without extensive supplementary analysis and we will explain why, after analyzing Table 7.1 by this method.

We recall that R_i is the average of all elements in row i and C_j the average of all elements in column j. There are p rows and q columns, and M represents the average of all elements in the table. In our case, $p = 10$ and $q = 5$. R_i and C_j are listed in the margins of Table 7.1.

The analysis of variance model is

$$y_{ij} = M + (R_i - M) + (C_j - M) + d^*_{ij} \tag{7.3}$$

In this equation, d^*_{ij} is called the row by column *interaction,* or residual. It is defined by

$$d^*_{ij} \equiv y_{ij} + M - R_i - C_j$$

Replacing d^*_{ij} in Eq. (7.3) by this expression reduces Eq. (7.3) to an identity.

$R_i - M$ is the *main row effect* for row i and $C_j - M$ is the *main column effect* for column j. If we write the preceding equation in the form

$$y_{ij} - M = (R_i - M) + (C_j - M) + d^*_{ij}$$

and then sum the squares of both sides over i and j, we can easily prove that the cross-product terms are all zero, so that we obtain the identity

$$\sum_i \sum_j (y_{ij} - M)^2 = q \sum_i (R_i - M)^2 + p \sum_j (C_j - M)^2 + \sum_i \sum_j d^{*2}_{ij}$$

Table 7.2 Analysis of Variance of Data of Table 7.1

Source	DF	SS	MS[a]
Rows (trees)	9	4,751,625	527,958
Columns (temperature)	4	$1{,}2606 \times 10^8$	31,515,370
$R \times C$	36	2.093,760	58,160

[a]MS = mean square = SS/DF

The four quantities appearing in this equation are called, respectively, the *total sum of squares* (left-hand side), the *row sum of squares,* the *column sum of squares,* and the *interaction (or residual) sum of squares*. Representing them by SS_{Total}, SS_{Rows}, $SS_{Columns}$, and $SS_{R \times C}$, it can easily be shown that

$$SS_{Total} = \sum_i \sum_j y_{ij}^2 - p \cdot q \cdot M^2$$

$$SS_{Rows} = q \sum_i R_i^2 - p \cdot q \cdot M^2$$

$$SS_{Columns} = p \cdot \sum_j C_j^2 - p \cdot q \cdot M^2$$

$$SS_{R \times C} = \sum_i \sum_j y_{ij}^2 - q \cdot \sum_i R_i^2 - p \cdot \sum_j C_j^2 + p \cdot q \cdot M^2$$

It is readily verified that

$$SS_{Total} = SS_{Rows} + SS_{Columns} + SS_{R \times C} \tag{7.4}$$

Each sum of squares corresponds to a number of degrees of freedom (DF), and a mean square (MS), which is SS/DF. We construct an analysis of variance table for the data of Table 7.1 and display it in Table 7.2.

Equation (7.4) is an identity and holds for *any* set of data. For real data, it is useful in that it *partitions* the total sum of squares into components that often have physical meaning. Thus, for our data, SS_{Rows} relates to differences between trees; $SS_{Columns}$ represents a squared *average* of the effect of temperature, and $SS_{R \times C}$ is what is left. It is apparent that the effect of temperature is large.

It can be shown that in the case of *additivity,* to be discussed in Section 7.4, the mean square corresponding to $SS_{R \times C}$, that is, to the interaction, is a measure of the variance of ε (see Eq. 7.2). In that case the classical analysis of variance is meaningful. But even then, it fails to show any detail in the analysis of the data. This will become clearer after we explore the data in detail, as we now do.

7.3 INTERNAL STRUCTURE, ROW AND COLUMN LINEARITY

A glance at Table 7.1 shows that the y_{ij} data possess an internal structure, quite apart from their relation to the two factors, trees and temperature. We see that the values in the body of the table (i.e., the y_{ij} values) decrease from left to right. This is of course due to the fact that the temperature values appear in an ordered way in the table but we can show that the internal structure of the y_{ij} is inherent in these values and is maintained even if we "mix up" the rows and/or the columns. To make this more apparent, we first introduce the following nomenclature: the x_1 and x_2 values are called "marginal labels" (in our case, trees and temperature). The values of y_{ij} are the "body" of the table.

It is seen that, had we interchanged some of the columns, for example columns 1 and 3, in Table 7.1, we could easily reestablish the original order, by means of the column averages without using the marginal labels. Our first objective now is to try to find the nature of the relation, if any, of *each row* to the corresponding C_j, or the relation of *each column* to the corresponding R_i. As an example, we have reproduced, in Table 7.3, the first row of Table 7.1, and the corresponding C_j. A plot of these data is shown in Fig. 7.1. Since no indication of curvature appears, we are encouraged to repeat the fit for all remaining rows of the table. The results are tabulated in Table 7.4a. Observe that one of the two parameters of each straight line, the height at the centroid (see Section 5.1) is simply the corresponding value R_i. The ten straight lines may be represented by the equation:

$$y_{ij} = R_i + B_i(C_j - M) + d_{ij} \tag{7.5}$$

Table 7.3 First Row of Table 7.1 Versus C_j

C_j	First row of Table 7.1
7568	6570
6773	6230
5253	4715
4210	3815
3259	3170

where B_i is the slope of the ith line and d_{ij} is the residual corresponding to y_{ij}. Note that the residual d_{ij} is what is left after fitting straight lines for each row; it is different from the residual d^*_{ij} occurring in Eq. (7.3) (Mandel, 1961).

Table 7.4b is a tabulation of the residuals d_{ij}. It can be proven and verified in Table 7.4b that the sum of the residuals in each row or in each column is zero (any difference from zero is the result of rounding errors). Furthermore, if we sum both sides of Eq. (7.5) over i, we obtain

$$\sum_i y_{ij} = \sum_i R_i + (C_j - M)\sum_i B_i$$

or

$$pC_j = pM + (C_j - M)\sum_i B_i$$

hence

$$\sum_i B_i = \frac{p(C_j - M)}{C_j - M} = p$$

or, for the average $\overline{B}$ of the B_i,

$$\overline{B} = \frac{p}{p} = 1$$

Table 7.4a lists, for each row, all the parameters occurring in Eq. (7.5): R_i, B_i, C_j, M. Note that the analysis so far has made no

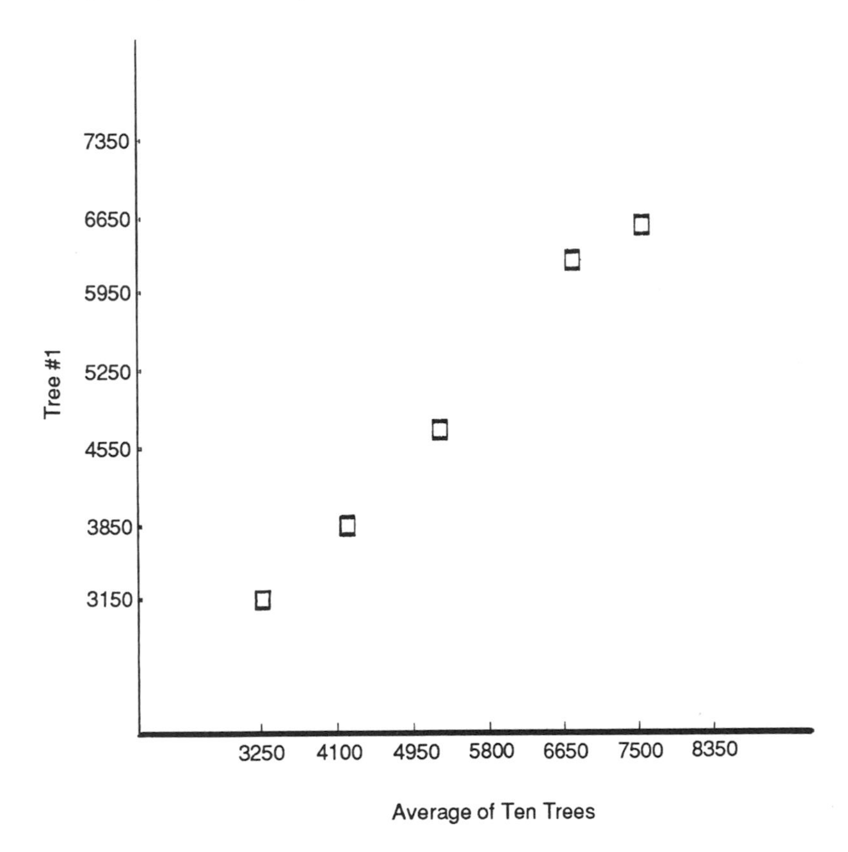

Figure 7.1 Compressive strength of timber tree number 1 versus average of ten trees.

use whatsoever of the marginal labels. It is strictly an internal structure analysis, summarized in Eq. (7.5) together with Tables 7.4a and 7.4b. The relative smallness and apparent randomness (see succession of algebraic signs in each row) of the residuals in each row of Table 7.4b shows that our attempt to express the rows as *linear functions* of the column averages C_j (see Eq. 7.5) was quite successful. The model we have fitted, as represented by Eq. (7.5), is called *row-linear,* for obvious reasons.

Table 7.4a suggests that the ten slope values are not identical. Of course, one would expect differences due to random errors in the

Table 7.4a Parameters of Row-Linear Fit for Data of Table 7.2

Row	Height (R)	Slope (B)	Standard deviation of fit	Standard error of slope	Column	C_j
1	4900	.8305	150	.0424	1	7568
2	5814	.9721	106	.0299	2	6773
3	4967	.8240	53	.0149	3	5253
4	5485	1.0012	183	.0514	4	4210
5	5433	.9885	188	.0530	5	3259
6	5454	.9960	85	.0239		
7	5312	1.0239	93	.0261		
8	5360	1.0486	51	.0143	M = 5412	
9	5967	1.2107	210	.0592		
10	5429	1.1045	174	.0489		
	Std. Dev. =	0.1149				

data, but that a systematic effect is involved is proved by the fact that the standard deviation between these slopes, 0.1149, is appreciably larger than the standard errors of the individual slopes, which are all of the order of 0.06 or less. We conclude that the slope-differences in Table 7.4a are real.

Of course, not every two-way table is row-linear. If not, one may

Table 7.4b Residuals From Row-Linear Fit for Data of Table 7.1

−120	200	−52	−86	58
41	−81	−9	135	−85
−48	72	−10	−26	13
112	−7	−160	−146	201
201	−198	−130	91	36
29	11	−120	79	1
11	−80	131	−31	−32
−15	−16	37	51	−57
−127	1	196	159	−230
−85	98	117	−226	95

attempt to express the *columns* of the table as linear functions of the *row-averages*. Such a model would be called *column-linear*. If that too fails to yield satisfactory linear relations, the internal structure is more complicated. It may then be advisable to attempt a transformation of scale, and try to fit straight lines in this new scale. In some cases, a logarithmic transformation will yield linearity.

We will continue the analysis of our data in Section 7.6, but digress at this point to discuss two important concepts: additivity and concurrence.

7.4 ADDITIVITY

If the slopes in our example had differed from each other by no more than their standard errors, then they would be essentially all equal to unity (their average) and the internal structure of Table 7.1 would be called *additive*.

The reason for the word additivity can readily be understood from Eq. 7.5. Indeed, if all B_i are equal, then they are all equal to unity: $B_i = 1$ for all i, and Eq. (7.5) becomes

$$y_{ij} = R_i + C_j - M + d_{ij} \tag{7.6}$$

which shows that y_{ij} is obtained, apart from the constant $-M$ and the random error d_{ij}, by *adding* the row function R_i to the column function C_j. The fact that in our case B_i varies appreciably from row to row precludes additivity. Our structure is therefore definitely *nonadditive*. The geometrical equivalent of additivity is *parallelism,* because the straight lines representing the individual trees in our analysis would be parallel to each other.

We need not discuss the additive situation in more detail, since it is simply a *special case* of the more general model expressed by Eq. (7.5). We have referred to this more general situation as the "row-linear model for two-way data." Additivity in the special case of the linear model is obtained when all $B_i = 1$.

7.5 CONCURRENCE

A close examination of Table 7.4a shows that B_i tends to increase systematically as R_i increases. The impression is confirmed by a graph

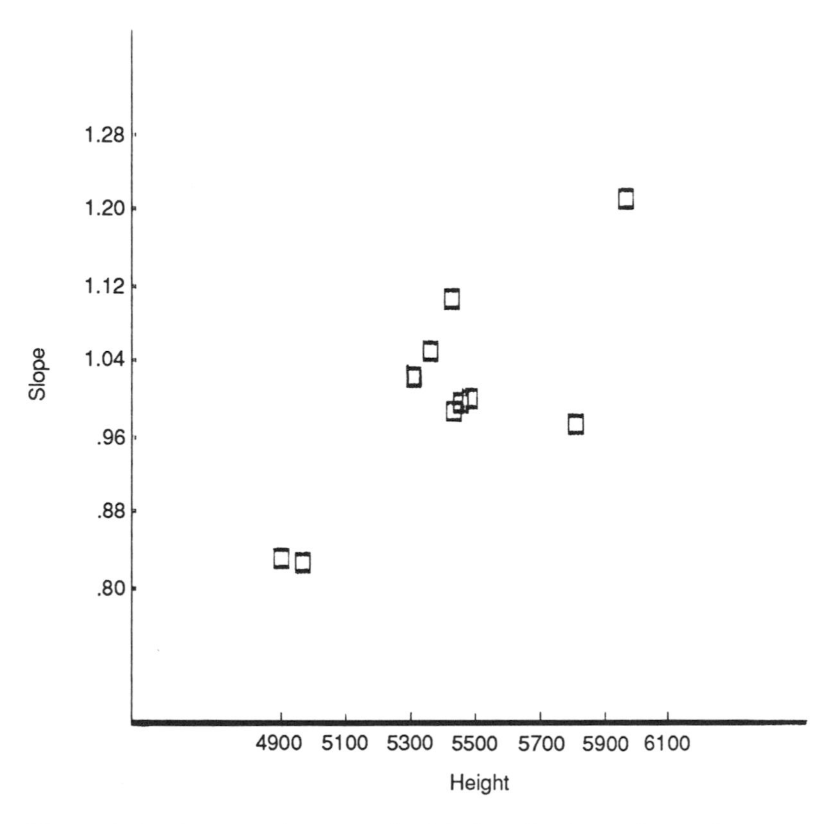

Figure 7.2 Compressive strength of timber. Deviation from column average $y_{ij}C_j$ versus column average C_j.

obtained by plotting, in Table 7.4a, the slopes of B_i versus the heights, R_i. This plot is shown in Fig. 7.2. The relationship is approximately linear. We propose to explore the meaning of such a relationship.

First, we observe that Eq. (7.5) is actually an expression for a *family* of straight lines, one for each row of the table (i.e., for each i value). More generally, Eq. (7.2) can be considered as representing a *family of curves,* one for each row of the table. Indeed, by selecting a specific row, by making x_1 a constant, say, $x_1 = C$, Eq. (7.2) becomes $y_{ij} = F(C,x_2)$, but since C is a constant, this is the equation of a curve of y versus x_2.

This family of lines is plotted in Figure 7.3, where the common

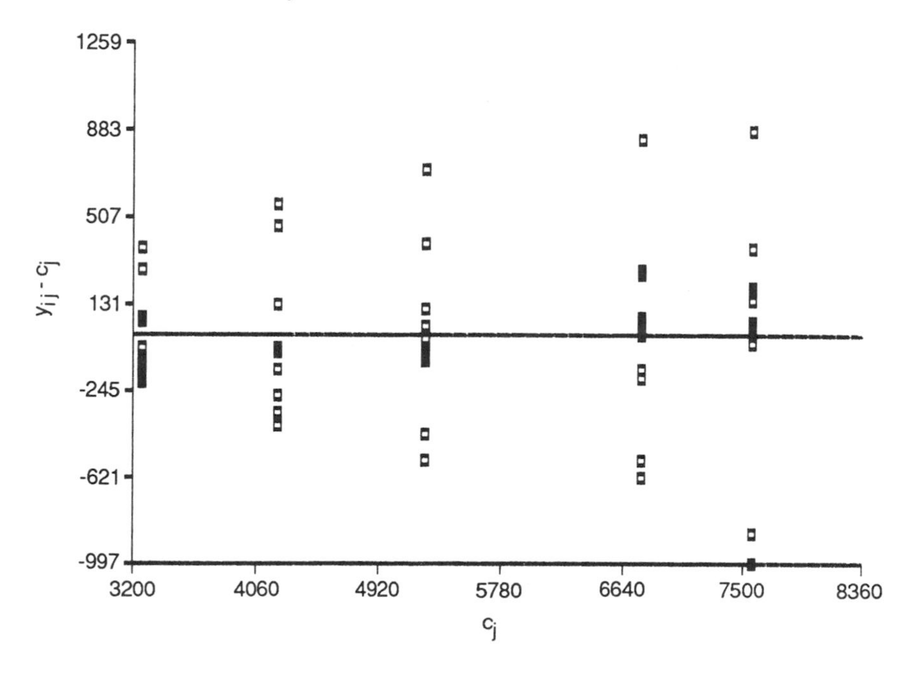

Figure 7.3 Compressive strength of timber slope versus height.

abscissa for all lines is C_j, and the ordinates for the points of the line for row i are $y_{ij} - C_j$.

Figure 7.3 suggests that all ten lines tend towards a common point, situated on the left side of the graph. Then a plot of y_{ij} versus C_j would also show this tendency for, if for some j, say j_0, $y_{ij} - C_j$ is a constant, say K, then $y_{ij_0} = C_{j_0} + K =$ constant. This intuitive insight can be expressed more rigorously by simple algebra, as we now show.

Suppose that, in accordance with Fig. 7.3, a straight line relationship exists between B_i and R_i. Then we have:

$$B_i = P + Q \cdot R_i \tag{7.7}$$

Since $\overline{B} = 1$ and $\overline{R} = M$, we have

$$1 = P + QM \qquad \text{and} \qquad B_i = 1 + Q(R_i - M)$$

Introducing this relation into Eq. (7.5), we obtain

$$y_{ij} = R_i + [1 + Q(R_i - M)](C_j - M) + d_{ij} \tag{7.8}$$

Let us subtract $M - 1/Q$ from both sides of Eq. (7.8):

$$y_{ij} - \left(M - \frac{1}{Q}\right) = \left(R_i - M\right) + \frac{1}{Q} + \left(C_j - M\right) + Q\left(R_i - M\right)\left(C_j - M\right) + d_{ij}$$

Represent $R_i - M$ by a_i, and $C_j - M$ by b_j

$$a_i = R_i - M \qquad b_j = C_j - M \tag{7.9}$$

$$y_{ij} - \left(M - \frac{1}{Q}\right) = a_i + \frac{1}{Q} + b_j + Qa_ib_j + d_{ij}$$

$$= a_i\left(1 + Qb_j\right) + \frac{1}{Q}\left(1 + Qb_j\right) + d_{ij}$$

or

$$y_{ij} - \left(M - \frac{1}{Q}\right) = \frac{1}{Q}\left(1 + Qa_i\right)\left(1 + Qb_j\right) + d_{ij} \tag{7.10}$$

Thus the quantity $y_{ij} - (M - 1/Q)$ is essentially the product of the two factors, the first depending on i only, and the second on j only.

Let us represent by $\hat{y}_{ij}$ the estimated value of y_{ij}. If C_j is chosen so that

$$b_j = -\frac{1}{Q} \qquad \text{or} \qquad C_j = M - \frac{1}{Q}$$

then

$$\hat{y}_{ij} - \left(M - \frac{1}{Q}\right) = 0 \qquad \text{(see Eq. 7.10)}$$

But, we also have

$$C_j - \left(M - \frac{1}{Q}\right) = 0$$

Thus, for $C = M - 1/Q$, $\hat{y}_{ij}$ is also equal to $M - 1/Q$, hence independent of i. At that value of C_j, the lines for all i pass through the common point

$$C_j = M - \frac{1}{Q} \qquad y = M - \frac{1}{Q} \tag{7.11}$$

Therefore, if Eq. (7.7) holds, that is, if B_i is a linear function of R_i, then we have a set of concurrent lines. The point of concurrence lies on the 45 degree-line of the y_{ij} versus C_j plot, and has coordinates given by Eq. (7.11). Let us represent these coordinates by y_0. Then

$$y_0 = M - \frac{1}{Q} \tag{7.12}$$

and Eq. (7.10) becomes

$$y_{ij} - y_0 = \left(M - y_0\right)\left(1 + \frac{a_i}{M - y_0}\right)\left(1 + \frac{b_j}{M - y_0}\right)$$

$$= \frac{1}{M - y_0}(M - y_0 + R_i - M)(M - y_0 + C_j - M)$$

or

$$\hat{y}_{ij} = y_0 + \frac{(R_i - y_0)(C_j - y_0)}{M - y_0} \tag{7-13}$$

Thus, in a concurrent model, the estimated value of y_{ij} is equal to a constant, y_0, plus a product of two quantities, one depending only on the row-parameter R_i, and the other depending only on the column-parameter C_j. If the point of concurrence *coincides with the origin* of the coordinate axes, y_0 *becomes zero,* and the model for the measured value y becomes

$$y_{ij} = \frac{R_i C_j}{M} + d_{ij} \tag{7.14}$$

This model is called *multiplicative,* for obvious reasons. It is interesting to note that if a set of data is multiplicative, *and if* the residuals d_{ij} are small with respect to the quantity measured, we have

$$\begin{aligned}
\ln y_{ij} &= \ln\left[\frac{R_iC_j}{M} + d_{ij}\right] \\
&= \ln\left[\left(\frac{R_iC_j}{M}\right)\left(1 + \frac{d_{ij}}{R_iC_j/M}\right)\right] \\
&\cong \ln\frac{R_iC_j}{M} + \frac{d_{ij}}{R_iC_j/M}
\end{aligned}$$

so that, denoting by e_{ij} the quantity

$$e_{ij} = \frac{d_{ij}}{R_iC_j/M}$$

we have

$$\ln y_{ij} = -\ln M + \ln R_i + \ln C_j + e_{ij} \tag{7.15}$$

It is, of course, assumed that the quantities of which logarithms are to be taken are all positive. Equation (7.15) shows that if a set of two-way data conforms to the multiplicative model, *and* if the errors of measurement are sufficiently small, the logarithm of the measurement will be additive.

Let us emphasize that the logarithmic transformation will *not* produce additivity if the point of concurrence is different from zero.

Finally, it is interesting to note that the concurrent model provides a sound rationale for a test that was proposed by Tukey (Tukey, 1949) and that is known as *Tukey's One Degree of Freedom for Nonadditivity*. Actually, the model represented by Eq. (7.5) is a far more general way of expressing nonadditivity than Tukey's test. It contains Tukey's model as a special case. Thus, if one wishes to examine nonadditivity, the first step would be to try the general linear model of Eq. (7.5) and only then examine whether concurrence applies, as is done in this section.

7.6 STRENGTH OF TIMBER: EXAMINATION OF DATA FOR CONCURRENCE

Returning now to the data of Table 7.1, we have already found that they are definitely not additive. We have also seen (Fig. 7.2) that a plot of B_i versus R_i produces an approximate straight line, thus suggesting concurrence. The slope Q of that line is found to be .0002813 (using the formulas of Section 5.2). We then find, by means of Eq. (7.12) that the point of concurrence is

$$y_0 = M - \frac{1}{Q} = 5412 - 1/.0002813 = 1858$$

Applying Eq. (7.13) with this value of y_0, and calculating the residuals $y_{ij} - y_{ij}$, we obtain the values in Table 7.5. Since the plot in Fig. 7.2 is only approximately a straight line, the concurrence itself is only approximate. Therefore we may expect somewhat larger residuals from the concurrent model than from the row-linear model. This is indeed the case, as shown by a comparison of Tables 7.4b and 7.5. Whether one adopts the row-linear model or the concurrent one depends on circumstances known to the subject matter specialist. This individual must decide whether the concurrent fit is sufficient from a practical point of view. So far the entire analysis has been in terms of internal structure only, and has been carried out without making any use of the temperature values.

Table 7.5 Residuals from Concurrent Model Fit for Data of Table 7.1

−175	166	−48	−56	113
−263	−273	14	305	218
−157	3	−2	35	122
70	−33	−157	−123	243
164	−221	−127	112	73
−5	−10	−118	98	35
123	−9	123	−93	−144
121	70	27	−25	−193
−9	75	188	93	−347
130	234	101	−346	−120

7.7 FITTING THE INTERNAL STRUCTURE PARAMETER C_j TO TEMPERATURE

Table 7.4a contains the "structural parameters" M, C_j, R_i, and B_i. To complete our empirical fit, we must fit C_j versus the temperature. M is a constant, and R_i and B_i are functions of a random variable, trees, and can therefore not be fitted. Figure 7.4 presents a plot of C_j versus temperature. For all practical purposes the relation is a straight line, with intercept = 6531 and slope = −55.90. In view of this result, each row of the table can now be fitted to temperature, either directly by linear regression, or by replacing C_j in Eq. 7.5 or Eq. 7.13 by its expression as a function of temperature:

$$C_j = 6531 - 55.90 \cdot \text{temperature}$$

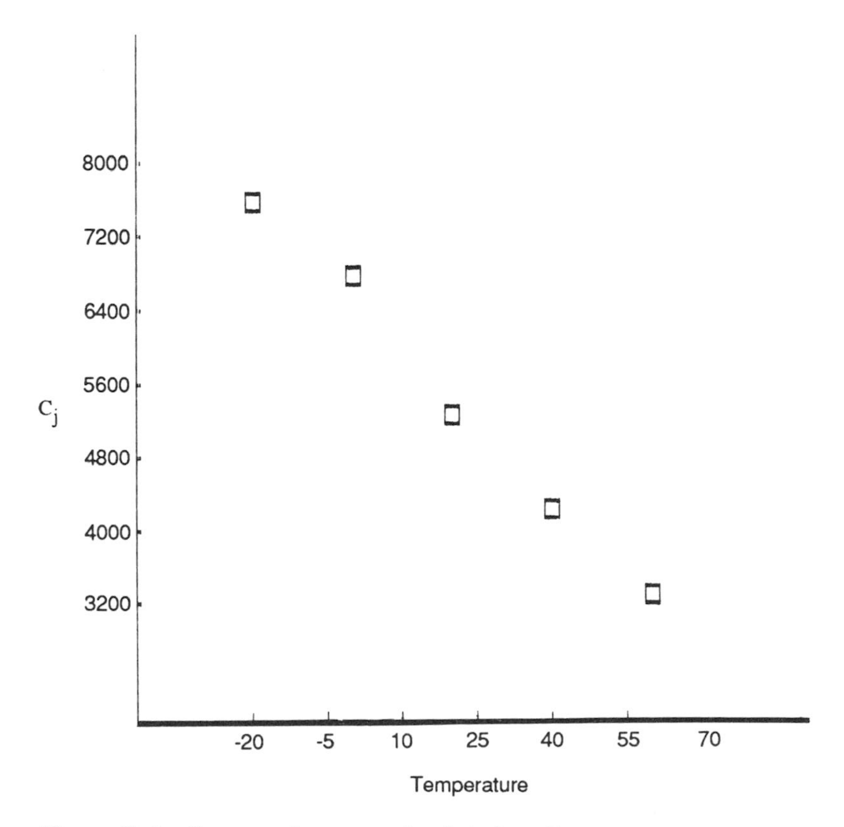

Figure 7.4 Compressive strength of timber C_j versus temperature.

Table 7.6 Residuals from Additive Fit for Data of Table 7.1

−485	−30	−25	118	424
−19	−119	−4	169	−25
−427	−167	18	186	392
115	−5	−160	−147	199
177	−213	−128	105	61
21	6	−119	84	10
63	−47	128	−59	−83
90	50	30	−7	−161
328	288	163	−94	−683
141	241	101	−351	−130

7.8 COMPARISON OF ADDITIVE AND ROW-LINEAR FITS

We have shown evidence that a row-linear model is a better fit than an additive one. Nevertheless, we wish to examine what we have achieved by fitting a row-linear model as against fitting an additive model. The quality of fit for the row-linear model is indicated by the residuals shown in Table 7.4b. Had we fitted an additive model we would then have obtained the residuals shown in Table 7.6 (see Eq. (7.3)). They are appreciably larger on average. Furthermore, some of the rows show noticeable trends, indicating systematic departures from additivity. This comparison confirms our preference for the row-linear model.

7.9 A DIFFERENT FORM FOR THE ROW-LINEAR MODEL

A table of data that conforms to the row-linear model satisfies the following equation:

$$y_{ij} = R_i + B_i(C_j - M) + d_{ij} \tag{7.16}$$

C_j is defined by

$$C_j = \sum_i \frac{y_{ij}}{p} \tag{7.17}$$

where p is the number of rows of the table (see Eq. 7.5), R_i is the average of row i, and M is the grand average of the table. The residual d_{ij} is assumed small with respect to y_{ij}. It is always possible to compute row- and column-averages and to write:

$$y_{ij} = M + (R_i - M) + (C_j - M) + d^*_{ij} \tag{7.18}$$

The residual d^*_{ij} represents what is left after fitting the three terms M, $R_i - M$, and $C_j - M$. The latter two are called the *effects* of row i and column j, respectively. Equation (7.18) represents an additive structure if and only if d^*_{ij} is a random experimental error. Equation (7.18) can be written in the form

$$y_{ij} = R_i + (C_j - M) + d^*_{ij}$$

$$= R_i + B_i(C_j - M) + [d^*_{ij} - (B_i - 1)(C_j - M)]$$

Comparing it to Eq. (7.16), we see that

$$d^*_{ij} = (B_i - 1)(C_j - M) + d_{ij} \tag{7.19}$$

Thus, the residual of the additive way of writing the model, d^*_{ij}, is equal to the sum of two terms: (a) d_{ij}, which is a random effect, assuming a row-linear model, and (b) a *systematic* term, $(B_i - 1)(C_j - M)$, which is the product of a row-dependent variable, $B_i - 1$, and a column-dependent variable, $C_j - M$. It follows that if a set of data is really row-linear, but we write it in its additive form (Eq. 7.18), then the residual d^*_{ij}, rather than being random, will actually contain a multiplicative systematic component, $(B_i - 1)(C_j - M)$.

7.10 THE ANALYSIS OF VARIANCE TABLE FOR LINEAR MODELS

The analysis of variance shown in Table 7.3 applies to additive models. The interaction term is then simply a measure for the error variance.

For a nonadditive structure, the interaction term is due to two sources: measurement error of the tabulated measurements, and non-additivity. It is possible to partition the interaction term into these two sources, when the structure is either row- or column-linear. We illustrate it for a row-linear model with p rows and q columns.

The basic equation for such a structure is

$$y_{ij} = M + (R_i - M) + (C_j - M) + (B_i - 1)(C_j - M) + \text{error}$$

Writing it as

$$y_{ij} - M = (R_i - M) + (C_j - M) + (B_i - 1)(C_j - M) + \text{error}$$

squaring both sides and adding over all i and j, we obtain

$$\sum_i \sum_j (y_{ij} - M)^2 = q \sum_i (R_i - M)^2 + p \sum_j (C_j - M)^2 + \sum_i (B_i - 1)^2 \sum_j (C_j - M)^2 + \text{error}$$

It is indeed readily proved that all cross-product terms are equal to zero. Thus, the usual interaction term which is $\Sigma_i \Sigma_j (y_i - R_i - C_j + M)^2$ is now partitioned into two terms: $\Sigma_i (B_i - 1)^2 \Sigma_j (C_j - M)^2$ and the error sum of squares. It can also be shown that the degrees of freedom corresponding to these two terms are $(p - 1)$, and $(p - 1)(q - 2)$. Table 7.7 is the analysis of variance table for the data of Table 7.1. We see that the partitioning has yielded two terms with very unequal mean squares. This tells us that it is worthwhile to consider a row-linear model. We emphasize here that the analysis of variance table by itself does not prove that row-linearity prevails, since a single observation, properly placed, can create very similar results in it. But having looked at the data in detail, as we did, this table is useful in providing quantitative results.

Table 7.7 Extended Analysis of Variance for Data of Table 7.1

Source	DF	SS	MS
Rows	9	4,751,625	527,958
Columns	4	1.2606×10^8	31,515,370
$R \times C$	36	2,093,760	58,160
Slopes	9	1,498,976	166,553
Residuals	27	594,784	22,029
Residual from concurrent	35	1,145,266	32,722

7.11 CONCLUSION

The usual analysis of variance model is based on an assumption of additivity. We have shown how additivity can be tested by fitting a row-linear (or column-linear) model and examining the slopes. Linearity of a plot of slopes versus row-averages in a row-linear model indicates a concurrent model. Final choice of a model depends on how close a fit one desires for practical use of the fit.

REFERENCES

Mandel, J. (1961). Non-Additivity in Two-Way Analysis of Variance, J. Am. Stat. Assoc., *56,* 878–888.

Tukey, J.W. (1949). One Degree of Freedom for Non-Additivity, Biometrics, *5*, 232–242

Williams, E.J. (1959). Regression Analysis, John Wiley & Sons, Inc., London.

8

A Fit of High-Precision Two-Way Data

8.1 SCOPE

In this chapter we apply the insights we have gained in the previous chapter to a set of very precise data. We also show how we can fit the structural parameters of the fit to the marginal labels of the table.

8.2 THE REFRACTIVE INDEX OF BENZENE

The data are shown in Table 8.1. They represent the quantity

$$y = \frac{n^2 - 1}{n^2 + 1} \tag{8.1}$$

where n is the refractive index of benzene.

The row label is pressure, expressed in atmospheres, and the column label is wavelength, expressed in angstrom. We also show R_i,

Table 8.1 Refractive Index of Benzene[a]

	Wavelength λ								
Pressure P	6678	6438	5876	5086	5016	4922	4800	4678	R_i
1	0.287528	0.288222	0.290225	0.294367	0.294870	0.295531	0.296490	0.297547	0.2930
246.2	0.293515	0.294242	0.296297	0.300495	0.301013	0.301708	0.302653	0.303744	0.2992
484.8	0.298498	0.299225	0.301279	0.305558	0.306072	0.306737	0.307739	0.308831	0.3042
757.2	0.303243	0.303965	0.306058	0.310383	0.310918	0.311597	0.312596	0.313714	0.3090
1107.7	0.308426	0.309152	0.311277	0.315688	0.316233	0.316921	0.317931	0.319058	0.3143
C_j	0.298242	0.298961	0.301027	0.305298	0.305821	0.306499	0.307482	0.308579	0.3039

[a]The tabulated value is $(n^2 - 1)/(n^2 + 2)$, where n = refractive index.

Table 8.2 Row-Linear Parameters for Data of Table 8.1

Column variables		Row variables			
Wavelength λ	Column average C	Pressure P	Row average R_i	Row slope B_i	Standard error of slope
6678	0.298242	1.	0.293098	0.9694	0.00032
6438	0.298961	246.2	0.299208	0.9885	0.00119
5876	0.301027	484.8	0.304242	0.9993	0.00071
5086	0.305298	757.2	0.309059	1.0133	0.00040
5016	0.305821	1107.7	0.314336	1.0301	0.00071
4922	0.306499		Stand. Dev.	0.0232	
4800	0.307482				
4678	0.308578				

the row averages; C_j, the column averages, and M, the grand average. Applying the row-linear model to these data (see Chapter 7), we obtain Table 8.2. This table exhibits row and column labels, the structural parameters of the row-linear model and the standard error of the slope for each row of the table. It also lists the standard deviation, .0232, between the slopes.

Figure 8.1 is a plot of $y_{ij} - C_j$ versus C_j for all rows. Table 8.3 lists the residuals, multiplied by 10^6, from the row-linear fit.

The slope versus height plot, obtained from the values shown in Table 8.2 is shown in Fig. 8.2. A straight line is a reasonable approximation for this relationship. Therefore, a concurrent fit seems to be appropriate. However, we will fit the parameters of the row-linear fit, rather than those of the concurrent fit, to the labels, pressure and wavelength.

8.3 EMPIRICAL FIT OF THE STRUCTURAL PARAMETERS

The fits in question can be verified to be nonlinear. They have to be fitted by nonlinear equations, the simplest of which is the quadratic equation

$$y = a + bx + cx^2 \tag{8.2}$$

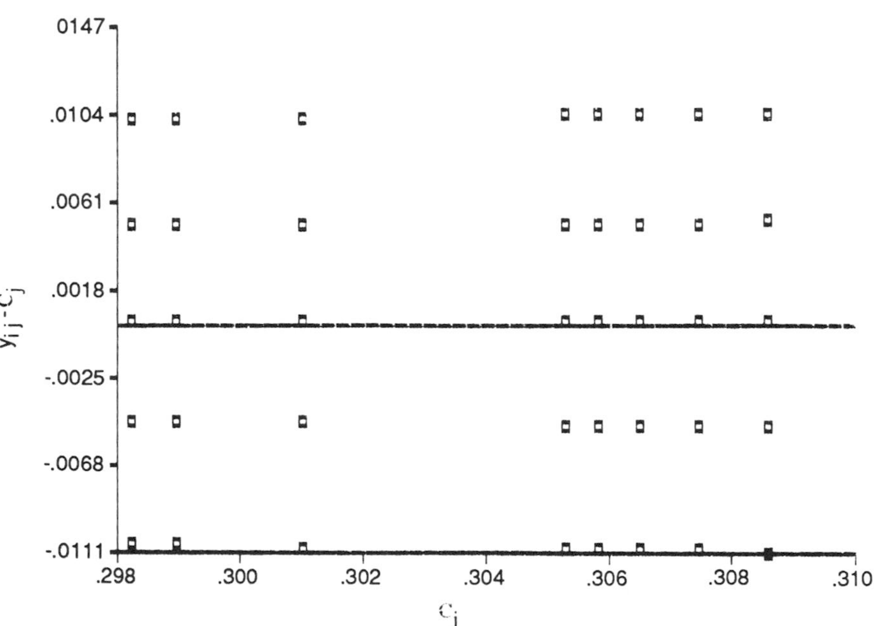

Figure 8.1 Refractive index of benzene. Deviation from column average $Y_{ij} - C_j$ versus column average C_j.

In many cases, Eq. (8.2) is appreciably better than the linear fit $y = a + bx$. However, for very precise data, it is often still inadequate, as we explain in the next section.

8.4 A GENERALIZED QUADRATIC EQUATION (QFP)

The quadratic, Eq. (8.2) always represents a curve composed of two symmetrical parts, one being the mirror image of the other, the "mirror" being a vertical line. Indeed, the y values obtained for $x_1 = -b/2c + \Delta$ and for $x_2 = -b/2c - \Delta$ are always identical, so that the vertical line at $x_0 = -b/(2c)$ is the line of symmetry. This property of symmetry is often an undesirable restriction, since many relations between experimental data are asymmetrical. Therefore, a generalization of the

Table 8.3 Residuals[a] (Data of Table 8.1) from Row-Linear Fit

4.349	0.608	−0.170	−0.979	−5.055	−0.892	4.898	−2.461
−11.705	4.683	16.660	−8.219	−7.094	17.562	−8.896	−2.707
−0.580	7.575	−3.233	6.787	−1.730	−14.137	4.963	0.659
7.190	0.197	−0.278	−2.988	1.730	−5.718	−2.920	3.153
5.959	−8.534	−10.229	4.354	10.668	1.209	−0.876	−2.526

[a]Multiplied by 10^6.

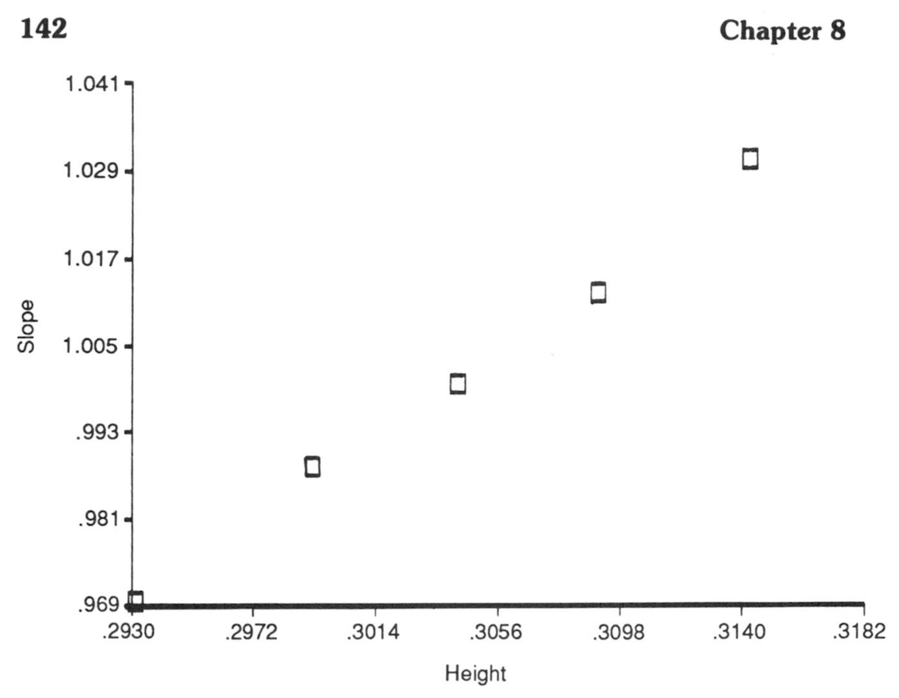

Figure 8.2 Refractive index of benzene. Slope versus height.

quadratic is indicated. We obtain it by using the Box-Cox Transformation (Box and Cox, 1964). This is a function of the type

$$z = \frac{x^{\alpha} - 1}{\alpha} \tag{8.3}$$

where α is a parameter whose value must be found to fit the particular application. Simply put, z is a linear function of the α-power of x. The reason for subtracting unity and dividing by α is the easily proven fact that

$$\lim_{\alpha \to 0} \frac{x^{\alpha} - 1}{\alpha} = \ln x \tag{8.4}$$

Thus, the family of functions represented by Eq. (8.3) includes, as one of its members, the logarithmic function, obtained in practice by selecting a value of α very close to zero, such as 0.001 or 0.0001 (making α exactly equal to zero will give an undetermined quantity, and, if used on the computer, signal an error).

We now generalize the quadratic of Eq. (8.2) by writing

$$y = a + b\left(\frac{x^{\alpha} - 1}{\alpha}\right) + c\left(\frac{x^{\alpha} - 1}{\alpha}\right)^2 \tag{8.5}$$

The advantage of using this function is that it can represent asymmetrical as well as symmetrical curves. It is far more flexible than Eq. (8.2). The price we pay for this greater flexibility is the introduction of an additional parameter, α, whose "best" value must be calculated.

Given a set of (x,y) pairs, the fitting of Eq. (8.5) can proceed as follows:

1. The first step is to normalize both x and y, as follows:

$$\text{For } x\text{: } u = 2 + \frac{x - \bar{x}}{\sqrt{\Sigma (x - \bar{x})^2}} \tag{8.6}$$

$$\text{For } y\text{: } v = \frac{y - \bar{y}}{\sqrt{\Sigma (y - \bar{y})^2}} \tag{8.7}$$

The v-quantity is seen to be a simple "standardization" of y. The average of v is zero, and its sum of squares is unity; v always lies in the interval -1 to $+1$.

The reason for using a slightly different normalization for x, by adding 2 to the simple standardized form, is as follows. The independent variable, x, in Eq. (8.5) is raised to the power α. To make this possible and unambiguous, x must be positive over its entire range of variation. Since u is to replace x, it must be positive; u as defined by Eq. (8.6) always lies in the range 1 to 3, and therefore satisfies this requirement.

2. The most appropriate value for α is found by a scanning process, easily carried out on the computer. It is often sufficient to cover the range $-2 < \alpha < +2$. Twenty values of α, equal to -1.8, -1.6, . . . , 0.0001, . . . , 1.8, 2.0 are "tried." The middle value, which would be zero, is seen to be replaced by 0.0001, for reasons given above. For each of these values of α, a simple quadratic fit is carried out by setting

$$z = \frac{u^{\alpha} - 1}{\alpha} \tag{8.8}$$

and then fitting

$$\hat{v} = a + b \cdot z + c \cdot z^2 \tag{8.9}$$

Calculate the sum of squares of residuals, $\Sigma\,(v - \hat{v})^2$, and select the value of α that yields the smallest sum of squares. Let us denote this value of α by α_0. The process is repeated by now trying a new set of 20 values in a much smaller range, say from $\alpha_0 - 0.2$ to $\alpha_0 + 0.2$. The value yielded by this second scanning, say $\hat{\alpha}$, is the one finally adopted.

As indicated above, the process is readily carried out on a computer, but would of course be prohibitively tedious without it.

3. Now, using $\alpha = \hat{\alpha}$, calculate z for each u by Eq. (8.8), and fit $\hat{v}$ by Eq. (8.9), thus obtaining estimates for a, b, and c. Let $\sqrt{\Sigma\,(y - \bar{y})^2}$ be denoted by S_y. Then Eq. (8.7) becomes: $y = \bar{y} + S_y \cdot v$ and Eq. (8.9) then yields:

$$\hat{y} = (\bar{y} + a \cdot S_y) + (b \cdot S_y) \cdot z + (c \cdot S_y) \cdot z^2$$

Hence, we obtain

$$\hat{y} = A + B \cdot z + C \cdot z^2 \tag{8.10}$$

where

$$A = \bar{y} + a \cdot S_y \tag{8.11a}$$

$$B = b \cdot S_y \tag{8.11b}$$

$$C = c \cdot S_y \tag{8.11c}$$

The method we have just outlined is called the *Quadratic Four Parameter* (QFP) curve fitting procedure (Mandel, 1981). We will illustrate it here by fitting the row-averages of the refractive index data (R_i in Table 8.2) to the corresponding pressure value (P in Table 8.2).

Table 8.4 shows the main features of the fitting process. The fit is obviously excellent, as can be seen by comparing $\hat{R}$ with R. The equations for the fit are as follows:

Table 8.4 QFP of R_i Versus Pressure, P (Table 8.2)

Data		Normalized data[a]				
P	R	u	v	z^b	$\hat{v}^c$	$\hat{R}^d$
1.	0.293098	1.400314	−0.657590	0.336702	−0.657964	0.293092
246.2	0.299208	1.683973	−0.288673	0.521169	−0.287368	0.299230
484.8	0.304242	1.959996	0.015276	0.672965	0.013890	0.304219
757.2	0.309059	2.275121	0.306122	0.822067	0.306550	0.309066
1107.7	0.314336	2.680596	0.624743	0.986088	0.624770	0.314336

[a] α-value found for (u,v) data $= 0.0001$: (from scanning procedure for α).

[b] $z = (u^\alpha - 1)/\alpha$: Box-Cox transformation.

[c] Least squares quadratic fit of v on z. $\hat{v} = a + bz + c \cdot z^2$; $a = -1.347119$, $b = 2.071186$, $c = -0.0724848$.

[d] $\hat{R} = 0.303989 + 0.16562 \times \hat{v}$ (from relation between R and v).

Coefficients for $\hat{R} = A + B \cdot z + C \cdot z^2$, $A = 0.281787$, $B = 0.0339641$, $C = -9.64767 \times 10^{-4}$. Standard deviation of fit $[\Sigma (R - \hat{R})^2/(5 - 4)]^{1/2} = 33 \times 10^{-6}$.

$$u = 2 + (P - 519.38)/864.4186$$

$$v = (R - 0.303989)/0.016562$$

$$z = (u^{0.0001} - 1)/.0001$$

$$\hat{v} = -1.347119 + 2.071186 \cdot z - 0.0724848 \cdot z^2$$

$$\hat{R} = 0.303989 + 0.016562 \cdot \hat{v}$$

It is worth noting that since α is very close to zero, z is very close to $\ln(u)$, and for this particular example, a simple quadratic fit (Eq. 8.2) on $\ln(u)$ would have been satisfactory, but this is by no means generally the case. Having obtained the coefficients for the fit, it is important to ascertain "how well one has done" by calculating the standard deviation of fit. This is accomplished by calculating the residuals [$(R - \hat{R})$ for our example], summing their squares, dividing them by the appropriate degrees of freedom, and taking the square root of the quotient.

The number of degrees of freedom is $N - 4$, where N is the number of points. The subtraction of four reflects the fact that four parameters (α, A, B, C) were fitted.

8.5 COMPLETION OF THE EMPIRICAL FIT, FITTING THE MARGINAL LABELS

Applying QFP to our three sets of data, C_j versus λ, R_i versus P, and B versus P, we obtain the results shown in Table 8.5. The empirical fit has now been completed.

Using a computer, one can now readily calculate a fitted ("predicted") value for any given pair of P and λ. This is done by calculating $\hat{c}$, $\hat{R}$, and $\hat{B}$ from the QFP fits and then introducing these values into

$$\hat{y} = \hat{R} + \hat{B}(\hat{C} - M)$$

where the constant M is the grand average of all values in the body of Table 8.1 ($M = 0.303980$). This has been done for all 40 (P,λ) combinations used in the original experiment with the results shown in Table 8.6. This table lists the differences (multiplied by 10^6) between

Table 8.5 Fits to Marginal Labels (Values from Table 8.1)

C versus λ

$$C = 0.330513 - 0.059886 \cdot Z - 0.019351 \cdot Z^2$$
$$\text{where } Z = (u^{-1.72} - 1)/(-1.72)$$
$$\text{and } u = 2 + [\lambda - 5436.75)/2069.377]$$

R versus P

$$R - 0.281787 + 0.0339641 + Z - 0.0009648 \cdot Z^2$$
$$\text{where } Z = (u^{0.001} - 1)/(0.0001)$$
$$\text{and } u = 2 + [(P - 519.38)/864.419]$$

B versus P

$$B = 0.941198 + 0.086248 \cdot Z - 0.002328 + Z^2$$
$$\text{where } Z = (u^{0.12} - 1)/(0.12)$$
$$\text{and } u = 2 + [(P - 519.38)/864.419]$$

the observed and the calculated values. They are larger than those based on the internal fit only (Table 8.3) but still constitute a good fit, of the order of 0.5 to 50 units in the sixth place. Their greatest usefulness, however, is not in being able to reproduce the original measurements, but rather for the purpose of predicting a value for any *new* combination of P and λ. Of course, the new values for P and λ must lie in the ranges of those upon which the fit is based: P in the range 1 to 1100, and λ in the range 4678 to 6678. Outside these ranges, the procedure would constitute an extrapolation. Since the fit is purely empirical, extrapolation cannot be justified.

Table 8.6 Residuals[a] from Complete Fitted Process (Analysis of Table 8.1)

116	116	112	64	84	59	57	67
83	105	119	57	84	81	49	75
163	174	160	120	136	94	104	117
89	86	83	31	60	23	18	42
117	108	105	75	107	68	59	76

[a]Multiplied by 10^6.

8.6 SUMMARY

We have shown that the row-linear model discussed in Chapter 7, combined with empirical curve fitting based on a modified quadratic equation (using the Box-Cox transformation) can yield excellent results for data of high precision.

REFERENCES

Box, G. E. P., and D. R. Cox (1964). An Analysis of Transformations. J. Roy. Stat. Soc., Ser. B, *26,* 211–252.

Mandel, J. (1981). Fitting Curves and Surfaces with Monotonic and Non-Monotonic Four Parameter Equations. J. Res. National Bureau of Standards, *86,* 1–25.

9

A General Treatment of Two-Way Structured Data

9.1 INTRODUCTION

In Chapter 7, we presented a method, based on a linear model, for analyzing data such as Table 7.1. While this method often succeeds in leading to a very usable empirical equation for the data, as we illustrated in Chapter 8, there exist cases that do not conform to this model. It is in fact easy to test whether a particular set of data does or does not obey a linear model. The test consists in examining whether at least one of the following conditions are fulfilled: (a) All rows are linear functions of the column averages, C_j; and (b) All columns are linear functions of the row averages R_i. If neither (a) nor (b) are true, then the linear model is not adequate.

In such cases, it is often possible to find a transformation of scale of the tabulated variable, in which (a) and/or (b) are true. In this chapter we follow a different approach. The great advantage of this approach is that it always leads to a complete representation of the data by a finite number of terms, and very often to a good approximation by a rather small number of terms.

9.2 BASIC MODEL (SINGULAR VALUE DECOMPOSITION)

The model we propose is as follows:

$$y_{ij} = A_iB_j + C_iD_j + E_iF_j + \cdots \quad (9.1)$$

Thus, the tabulated variable is a sum of products in which each term is the product of a "row-variable" (dependent on i only) by a "column-variable" (dependent on j only).

For reasons that will become clear as we proceed, the various functions in Eq. (9.1) are first standardized as follows:

A_i is written as k_1u_i

$$A_i = k_1u_i \quad (9.2)$$

where we stipulate

$$\sum_i u_i^2 = 1 \quad (9.3)$$

B_j is written as k_2v_j

$$B_j = k_2v_j \quad (9.4)$$

where we stipulate

$$\sum_j v_j^2 = 1 \quad (9.5)$$

Thus:

$$A_iB_j = (k_1k_2)u_iv_j$$

Instead of k_1k_2 we write θ, so that

$$A_iB_j = \theta u_i v_j \tag{9.6}$$

where u_i and v_j are subject to the constraints expressed by Eqs. (9.3) and (9.5). The quantity θ is by convention always positive (unless it is zero). The problem is to find θ, u_i, and v_j. Note that u_i and v_j are *vectors*, whereas θ is a constant.

Similar considerations apply to the terms C_iD_j, E_iF_j, etc., in Eq. (9.1), so that our model can be written

$$y_{ij} = \theta_1 u_{1i} v_{1j} + \theta_2 u_{2i} v_{2j} + \cdots \tag{9.7}$$

The unknown constants and vectors in Eq. (9.7) are subject to further mathematical conditions. Briefly, if the model contains only one term, $\theta_1 u_{1i} v_{1j}$, then these parameters must satisfy the condition that after this term is accounted for, the sum of squares of the residuals is minimum. Thus, for a model with one term:

$$y_{ij} = \theta_1 u_{1i} v_{1j} + d_{ij} \tag{9.8}$$

the condition is that

$$\sum_i \sum_j d_{ij}^2$$

be as small as possible.

If the model contains, say, two terms then the second term is part of d_{ij}, so that

$$d_{ij} = \theta_2 u_{2i} v_{2j} + d_{ij}^*$$

where, again, the sum of squares of d_{ij}^* is minimum. The vectors u_{1i}, u_{2i}, . . . , v_{1j}, v_{2j}, . . . are called *eigenvectors* and the squares of the constants θ_1, θ_2, . . . are called *eigenvalues* or, sometimes, *latent roots*.

We will not go into the theoretical aspects (see, e.g., Mandel, 1971), but we will present a simple procedure for calculating all parameters in Eq. (9.7). Equation (9.7) is known as the Singular Value Decomposition (SVD) of the original table of data.

Table 9.1 The Singular Value Decomposition: An Artificial Example

	Data				First	Second
	3.02	2.01	3.97	1.04	23.11	5.4773
	6.96	5.03	9.02	3.03	56.20	12.7889
	8.05	5.00	10.99	1.96	58.86	14.6195
Trial vector	1	2	3	4		
First	934.7672	623.4371	1245.5421	309.6860		
Stand	0.5480	0.3655	0.7302	0.1805		
Second	223.2392	148.4350	297.7691	73.1010		
Stand	0.5481	0.3645	0.7311	0.1795		

9.3 NUMERICAL PROCEDURE

The computations are explained in terms of a small example, shown in Table 9.1. While all steps are extremely simple, they are time consuming, so that a computer is practically indispensable.

1. Below the table, as an additional row, we write *any* set of numbers, for example 1, 2, 3, 4 (the numbers are arbitrary but should be different from each other). We now compute the inner product of this additional row with each row of the table. An *inner product* of two vectors with the same number of elements is the sum of the products of corresponding terms. Thus, for the first row of the table, we have the inner product:

$$(1 \times 3.02) + (2 \times 2.01) + (3 \times 3.97) + (4 \times 1.04) = 23.11$$

This number is written on the right, next to row 1. This process yields the three numbers 23.11, 56.20, and 58.86.

2. We form the inner product of the vector consisting of these three numbers with each *column* of the table, obtaining the numbers appearing below 1, 2, 3, 4.

3. This set of four numbers is now "standardized" as follows:

(a) Compute the square root of the sum of their squares; this yields 1705.7972.

(b) Divide each of the four numbers by this value. We thus obtain the vector (0.5480 0.3655 0.7302 0.1815).

4. The entire process is now repeated, using this vector, rather than the numbers 1, 2, 3, 4. We thus obtain the vector (0.5481 0.3645 0.7311 0.1795).

5. Repeat the entire process until no appreciable change is noted in this vector.

6. The vector which has been computed in this way is v_j. To obtain θ and u_i, proceed as follows:

7. Form the inner product of v_j with every row of the table. This produces the vector θu_i. To obtain u_i, simply standardize θu_i by dividing each element by the square root of the sum of the squares. In this case, we obtain for θu_i:

5.4770 12.7865 14.6214

The standardizing factor is

$$\sqrt{5.4770^2 + 12.7865^2 + 14.6214^2} = 20.1811, \text{ which } = \theta.$$

This completes the calculations for the first term.

8. To obtain the second term, we first computer the residuals, i.e., the difference

$$y_{ij} - \theta_1 u_{1i} v_{1j}$$

for each cell of the table (each i,j combination). Then the entire procedure is carried out again, using the residuals rather than the original table. The results are shown in Table 9.2.

9.4 PROPERTIES OF EIGENVECTORS AND EIGENVALUES

First we note that the process always terminates: the number of terms in the model of Eq. (9.7) is finite. It is called the *rank* of the original table, and is equal, at most, to the number of rows or number of columns, whichever is smaller. For our example, the rank is 3. All vectors u are orthogonal to each other. This means that the inner product

Table 9.2 Analysis of Table 9.1

First term	θ = 20.18113				
	u-vector = [	0.271394	0.633589	0.724507]	
	v-vector = [	0.548121	0.364438	0.731125	0.179456]
Second term	θ = 1.196748				
	u-vector = [	−0.054375	−0.741771	0.668445]	
	v-vector = [	0.049247	−0.413558	0.372754	−0.829216]
Third term	θ = 0.02805805				
	u-vector = [	0.960937	−0.220797	−0.166881]	
	v-vector = [	0.783393	−0.480502	−0.376486	0.116928]

of a *u* vector with any other *u* vector is zero. The same holds for the *v* vectors. The "singular values" θ_1, θ_2, θ_3, . . . , appear in descending order, the largest first. The iteration process we used for the computations is generally quite rapid. However, if two eigenvalues are close to each other, it may take many iterations before convergence is reached. As mentioned earlier, the process always works, and reproduces, after all terms are calculated, the *exact* value of the y_{ij} in all cells. For real data, one often observes a drastic change in the magnitude of the singular values after a small number of terms. If, for example, the third singular value is very much smaller than the second, then two terms may suffice to produce a good approximation to the data.

9.5 AN EXAMPLE: STUDENT'S *t* DISTRIBUTION

Table 9.3 is a 7 × 8 selected portion of the table of values of Student's *t*, for seven values of the degrees of freedom and eight values of probability. Table 9.4 gives the first three terms (we could compute as many as 7) of the SVD of Table 9.3. This table provides three θ-constants, three sets of u_i, and three sets of v_j. The residuals after three terms are very small, so that we may consider these three terms as describing very adequately the original table. In order to express *t* as a function of degrees of freedom and probability, it would be nec-

Table 9.3 Frequency Distribution of Student's t

	Probability (right-hand tail)							
DF	0.60	0.75	0.90	0.95	0.975	0.990	0.995	0.999
5	0.267	0.727	1.476	2.015	2.571	3.365	4.032	5.893
8	0.262	0.706	1.397	1.860	2.306	2.896	3.355	4.501
11	0.260	0.697	1.363	1.796	2.201	2.718	3.106	4.025
16	0.258	0.690	1.337	1.746	2.120	2.583	2.921	3.686
21	0.257	0.686	1.323	1.721	2.080	2.518	2.831	3.527
26	0.256	0.684	1.315	1.706	2.056	2.479	2.779	3.435
120	0.254	0.677	1.289	1.658	1.980	2.358	2.617	3.160

essary to relate the u-vectors to the degrees of freedom and the v-vectors to probability. We will not attempt to do this for this example.

9.6 RELATION TO LINEAR MODEL

In Chapter 8 we analyzed two-way data (Table 8.1) by a linear model. In the present chapter we present a technique based on the singular value decomposition of a matrix. It is pertinent to ask in what way the two procedures are connected.

Let us suppose that a particular table of data "obeys" the row-linear model exactly. Then, according to Section 7.9, the model can be written in the form

$$y_{ij} = M + (R_i - M) + (C_j - M) + d_{ij}^* \tag{9.9}$$

where

$$d_{ij}^* = (B_i - 1)(C_j - M) + d_{ij} \tag{9.10}$$

Since d_{ij} is just a random error, d_{ij}^* is essentially a multiplicative quantity: row variable × column variable. Therefore, it makes sense to subject it to an SVD analysis:

$$d_{ij}^* = \theta u_i v_j + \text{residual} \tag{9.11}$$

The residual in this equation should essentially be d_{ij}. Furthermore, comparing Eqs. (9.10) and (9.11), we see that v_j should be closely

Table 9.4 SVD of Table 9.3

1	$\theta_1 =$	17.6351							
	$u_{1i} =$	0.4916	0.4057	0.3749	0.3524	0.3416	0.3353	0.3161	
	$v_{1j} =$	0.0385	0.1035	0.2027	0.2675	0.3287	0.4080	0.4685	0.6173
2	$\theta_2 =$	1.0114							
	$u_{2i} =$	0.7696	0.1276	−0.0753	−0.2130	−0.2750	−0.3102	−0.4150	
	$v_{2j} =$	−0.0883	−0.2243	−0.3666	−0.3933	−0.3575	−0.2158	−0.0317	0.6910
3	$\theta_3 =$	0.0257							
	$u_{3i} =$	−0.3796	0.5168	0.4264	0.1548	−0.0105	−0.1263	−0.6058	
	$v_{3j} =$	−0.1597	−0.4258	−0.4382	−0.2257	0.0450	0.3810	0.5276	−0.3531

related to $C_j - M$, and that u_i is proportional to $B_i - 1$. It can indeed be shown that if a row-linear model is written in the additive form of Eq. (9.9), and the residuals are then analyzed by the SVD technique (Eq. 9.11), we must have:

$$v_j = K(C_j - M) \tag{9.12}$$

Since $\sum_j v_j^2 = 1$, it follows that

$$K = \frac{1}{\sqrt{\sum_j (C_j - M)^2}} \tag{9.13}$$

Combining Eqs. (9.9), (9.11), and (9.12), we obtain

$$\begin{aligned} y_{ij} &= M + (R_i - M) + (C_j - M) + \theta u_i K(C_j - M) + d_{ij} \\ &= M + (R_i - M) + (1 + \theta K u_i)(C_j - M) + d_{ij} \end{aligned}$$

or

$$y_{ij} = R_i + B_i(C_j - M) + d_{ij} \tag{9.14}$$

with

$$B_i = 1 + \theta K u_i \tag{9.15}$$

where K is defined by Eq. (9.13).

In summary, assume that a set of data conforms to a row-linear model:

$$y_{ij} = R_i + B_i(C_j - M) + d_{ij}^*$$

Fit the data first by the additive terms

$$M + (R_i - M) + (C - M)$$

and the residuals of this fit by the singular value decomposition: Then, defining K by $K = (\Sigma\,(C_j - M)^2)^{-1/2}$ and B_i by $B_i = 1 + \theta K u_i$, we should obtain results close to those derived from the row-linear fit. Let us now apply this technique to the residuals from the additive fit for the data of Table 8.1

We obtain after some calculations:

Residual = .00048666 $u_i v_j$.

where u_i = [−0.660378 −0.252102 −0.017160 0.280775 0.649012] and v_j = [−0.525627 −0.485438 −0.299268 0.132078 0.197334 0.225966 0.321974 0.432887]

The constant K is equal to 94.60895 (see Eq. 9.13). Thus, $1 + \theta K u_i$ becomes $1 + 0.04604\ u_i$. It is readily verified that the quantities $1 + 0.04604 u_i$ are very close to the B_i-values shown in Table 8.2

9.7 THE BIPLOT

Row- or column-linearity are frequently encountered in physical or chemical data sets. In 1978, R. K. Gabriel presented an interesting graphical technique (Gabriel and Bradu, 1978) for testing whether a particular set of data is either row or column linear, or both.

Consider a situation in which Eq. (9.7) contains essentially two terms. This means that the term $\theta_3 u_{3i} v_{3j}$ and all subsequent terms are small enough with respect to the first two terms to be considered an experimental error.

We write, approximately, omitting the error term:

$$y_{ij} = \theta_1 u_{1i} v_{1j} + \theta_2 u_{2i} v_{2j}$$

or, more simply:

$$y_{ij} = \theta_1 u_i v_j + \theta_2 u_i' v_j' \qquad (9.16)$$

Multiply both sides of Eq. (8.16) by u_i, and sum over i. This yields:

$$\sum_i y_{ij} u_i = \theta_1 v_j \sum_i u_i^2 + \theta^2 v_j' \sum_i u_i u_i'$$

Now, we know that

$$\sum_i u_i^2 = 1 \qquad \text{and} \qquad \sum_i u_i u_i' = 0$$

Therefore:

$$\sum_i y_{ij} u_i = \theta_1 v_j \qquad (9.17a)$$

Similarly, we obtain:

$$\sum_i y_{ij} u_i' = \theta_2 v_j' \tag{9.17b}$$

Let us now assume that y_{ij} has a row-linear structure. Then:

$$y_{ij} = R_i + B_i(C_j - M) + \text{error} \tag{9.18}$$

Introducing Eq. (9.18) into (9.17a) and (9.17b), we obtain, neglecting the error term:

$$\sum_i R_i u_i + (C_j - M) \sum_i B_i u_i = \theta_1 v_j \tag{9.18a}$$

and

$$\sum_i R_i u_i' + (C_j - M) \sum_i B_i u'^i = \theta_2 v_j' \tag{9.18b}$$

The summation terms over i yield constants. Therefore, Eqs. (9.18) can be written:

$$M_1 + (C_j - M)N_1 = \theta_1 v_j \tag{9.19a}$$

$$M_2 + (C_j - M)N_2 = \theta_2 v_j' \tag{9.19b}$$

where M_1, M_2, N_1, N_2 are constants. If we now eliminate $(C_j - M)$ between these two equations, we obtain a *linear relationship* between v_j and v_j':

$$M_1 N_2 - M_2 N_1 = (N_2 \theta_1) v_j - (N_1 \theta_2) v_j' \tag{9.20}$$

Thus, a plot of v_j' *versus* v_j *should yield a straight line*. We have proved that if y_{ij} has *row-linear* structure, then the eigenvectors v_j and v_j' are linearly related. We can similarly show that if y_{ij} has *column-linear* structure, then u_i and u_i' are linearly related. Gabriel has called a point (u_i, u_i') a "row-marker" and a point (v_j, v_j') a "column-marker." He has proposed a simple plot, the *biplot*, of row and column markers. If the row-markers exhibit a linear relationship, the y_{ij} are column-linear. If the column-markers exhibit a linear relationship, y_{ij} is row-linear.

As an illustration of the usefulness of the biplot, Fig. 9.1 exhibits the row markers (filled squares) and the column markers (open squares)

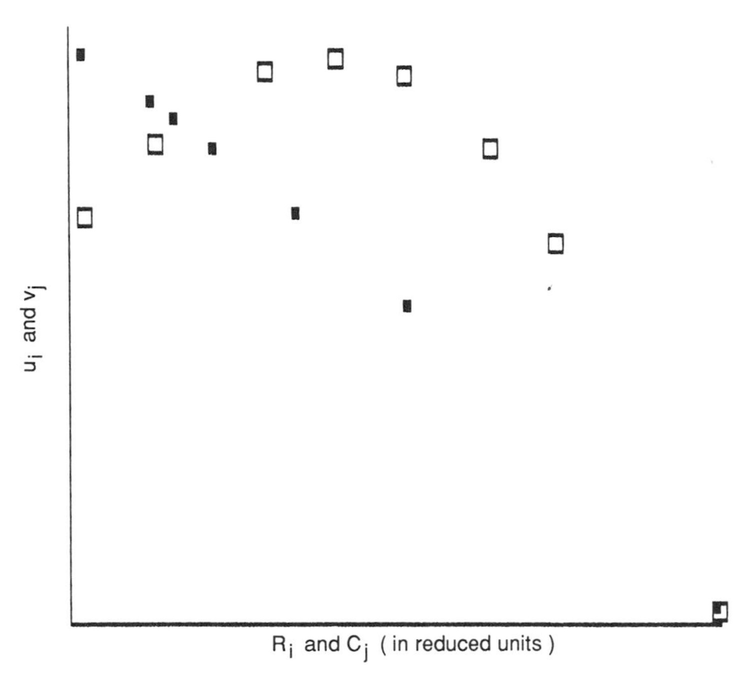

Figure 9.1. Biplot of *t*-distribution (Table 9.3). Plotting units are chosen so as to fill graph.

for the data of Table 9.3 (Student's *t* distribution). It is apparent that the row markers are close to, but not quite linear. Therefore the original data are close to, but not entirely column-linear. This can be checked by making a plot of all columns of Table 9.3 against the row-averages, as follows:

Let y_{ij} represent the entry in row i and column j. Then a column linear model requires:

$$y_{ij} = C_j + B_j(R_i - M) + \text{error}$$

Subtracting R_i from both sides and omitting the error term we obtain

$$y_{ij} - R_i = (C_j - B_jM) + (B_j - 1)R \qquad (9.21)$$

Figure 9.2 is a plot of $y_{ij} - R_i$ for all columns (j-values) against R_i. According to Eq. (9.21), the plot for each j is, except for experimental

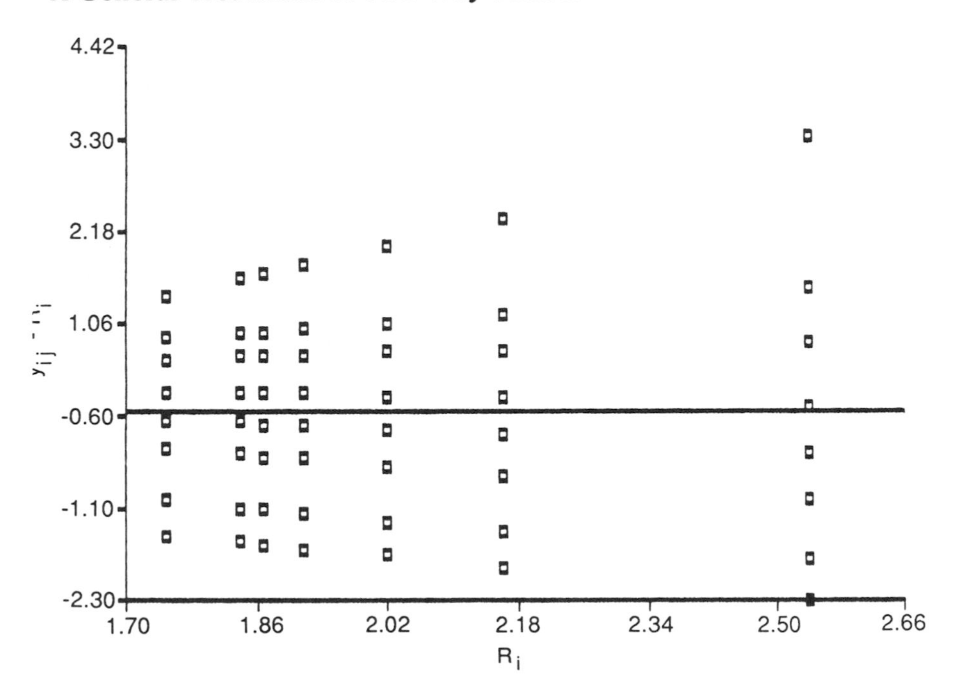

Figure 9.2. Student's *t*-distribution. Deviation from row average $y_{ij} - R_i$ versus row average R_i.

error, a straight line with slope $B_j - 1$. Figure 9.2, when carefully examined shows lines with slight curvature, in accordance with the curvature of the row markers in Fig. 9.1.

The biplot is conceived by Gabriel as a diagnostic tool. It allows us to recognize exact or approximate row or column linearity. It does not, by itself, provide quantitative parameters for the model.

9.8 ADDITIVE AND MULTIPLICATIVE TERMS

We have discussed row- and column-linear models. For any set of two-way data (data arranged by rows and columns) in which there are no missing values, there are four possibilities. The data are row-linear but not column-linear; column-linear but not row-linear; both row- and column-linear (concurrent); or neither row- nor column-linear.

The biplot allows us to make a simple diagnosis. Subsequently, we should make a more detailed analysis, as we did for the refractive index data in Chapter 8. If the data are neither row nor column linear, but exhibit other structural patterns, then sometimes a transformation of scale can create row or column linearity. Let us observe at this point that an additive model is really a concurrent one, with the point of concurrence at infinity. Thus, an additive structure is both row and column linear.

In making a biplot, it is often extremely useful to first code all data by subtracting a convenient constant, generally the grand mean, from all measurements. The properties of the matrix, in regard to row or column linearity, or concurrence, remain unchanged by such an operation, but the sensitivity of the plot to detect departures from linearity in the row or column markers can be greatly improved.

Let us observe, also, that data resulting from interlaboratory studies, in which the rows represent laboratories, are almost always row-linear. Often they are also concurrent.

We now present a model that incorporates both additive and multiplicative terms:

$$y_{ij} = A_i + B_j + (C_iD_j + E_iF_j + \cdots)$$

We gain additional insight in this model by starting with the usual analysis of variance model for two-way data (see Eq. 7.3a):

$$y_{ij} = M + (R_i - M) + (C_j - M) + \text{Interaction}$$

or

$$y_{ij} = R_i + C_j - M + \text{Interaction} \qquad (9.22)$$

We now apply the SVD analysis to the interaction term, and obtain:

$$y_{ij} = R_i + C_j - M + (\theta_1 u_{1i} v_{1j} + \theta_2 u_{2i} v_{2j} + \cdots) \qquad (9.23)$$

All terms are readily calculated: the additive terms R_i, C_j, G by the usual analysis of variance procedure, and the θ, u, and v terms from an SVD of the residuals from the additive model.

An example is given by the pentosans data (Table 1.2). The residuals from the additive terms (i.e., the values corresponding to the "interaction" term in Eq. (9.22)) are shown in Table 9.5. Their sum

Table 9.5 Residuals from Additive Terms Pentosans Data (Table 1.2)

−0.1070	0.1631	−0.1459	−0.2398	−0.1208	−0.1936	0.2169	0.1969	0.2302
0.0934	0.0801	0.0345	0.0672	0.0729	0.0101	0.1639	−0.3328	−0.1894
0.1493	0.0327	0.0804	0.1165	0.1255	0.0060	0.0365	−0.2402	−0.3068
−0.1218	−0.0917	−0.0407	−0.0380	−0.0789	−0.0751	−0.0846	0.2220	0.3087
0.1671	−0.0662	0.0415	0.0543	0.0433	0.4905	−0.1157	0.0576	−0.6724
0.3171	0.2938	0.3149	0.3209	0.3466	0.0238	0.0009	−0.5057	−1.1124
−0.4983	−0.4119	−0.2849	−0.2811	−0.3888	−0.2616	−0.2178	0.6022	1.7422

of squares is equal to 7.9040. If a SVD with a singular multiplicative term is carried out on these residuals, we obtain $\theta = 2.682152$. As a result of this operation, each term of Table 9.5 is approximated by $\theta u_i v_j$. The sum of squares of these terms, over all i and j, is:

$$\sum_i \sum_j \theta^2 u_i^2 v_j^2 = \theta^2 \sum_i u_i^2 \sum_j v_j^2 = \theta^2$$

since

$$\sum_i u_i^2 = \sum_j v_j^2 = 1$$

Thus, the sum of squares of the values estimated by SVD for the residuals is $(2.682152)^2 = 7.1939$. We have therefore accounted for practically the entire interaction (7.9040). The new residuals, *after* the $\theta u_i v_j$ term, have a sum of squares of $7.9040 - 7.1939 = 0.7101$, which is less than one-tenth of the original ones. Since there are $7 \times 9 = 63$ values, on the average a single new residual will be $\sqrt{0.7101/63} = 0.106$. The actual new residuals are shown in Table 9.6.

9.9 ANALYSIS OF VARIANCE FOR ADDITIVE AND MULTIPLICATIVE TERMS

It is of considerable value to construct an analysis of variance table similar to that shown in Table 7.2, but including not only the two additive terms (rows and columns), but also the multiplicative interaction terms $\theta u_i v_j$. The calculation of the *sum of squares* corresponding to these terms presents no problems. We have just shown that $\sum_i \sum_j (\theta u_i v_j)^2 = \theta^2$. This, then, is the sum of squares for a multiplicative term. But an analysis of variance table also requires the quantity we identified as *degrees of freedom*. How do we calculate the degrees of freedom corresponding to a multiplicative term? The answer is obtained from the following argument.

If we had analyzed a table of random normally distributed numbers of mean zero and variance σ^2, then *all* terms in the analysis of variance table should yield mean squares whose expected value is σ^2,

Table 9.6 Residuals from (Additive + One SVD Term) Pentosans Data (Table 1.2)

−0.0198	0.2242	−0.0881	−0.1787	−0.0482	−0.1442	0.2368	0.0838	−0.0656
0.0097	0.0214	−0.0210	0.0085	0.0032	−0.0374	0.1447	−0.2241	0.0949
0.0429	−0.0419	0.0099	0.0419	0.0370	−0.0543	0.0121	−0.1021	0.0544
−0.0198	−0.0202	0.0269	0.0336	0.0060	−0.0173	−0.0613	0.0897	−0.0376
0.0065	−0.1787	−0.0649	−0.0584	−0.0904	0.3994	−0.1525	0.2660	−0.1271
−0.0237	0.0550	0.0890	0.0819	0.0630	−0.1695	−0.0771	−0.0634	0.0448
0.0042	−0.0598	0.0481	0.0731	0.0294	0.0232	−0.1028	−0.0499	0.0362

Table 9.7 Analysis of Variance Pentosans Data (Table 1.2)

Source	DF	SS	MS
Rows	6	1.947	.3246
Columns	8	1637.08	204.64
R × C	48	7.9040	.1647
EV 1	20.95	7.1939	.3434
EV 2	12.75	.4321	.0339
EV 3	7.79	.2022	.0259
EV 4	6.51	.0759	.0116

since there is no structure in such data. Since MS = SS/DF, we would then have, for the term $\theta u_i v_j$:

$$\sigma^2 = \frac{E(\theta^2)}{DF}$$

where $E(\theta^2)$ is the expected value of θ^2. Hence

$$DF = \frac{E(\theta)^2}{\sigma^2} \tag{9.24}$$

Thus, DF would be known if we knew $E(\theta^2)$ for a table of random normal deviates of mean zero and variance σ^2. Monte Carlo calculations were performed on the computer to obtain this expected value: by constructing such a table with $\sigma^2 = 1$, thousands of times, and averaging θ^2 over all cases, a satisfactory approximation can be obtained for $E(\theta^2)$. Table A4 shows the results for various numbers of rows and columns. It can be shown (Mandel, 1971) that for a 7 × 9 table, the result for the first interaction term is the same as for a 6 × 8 table in which the SVD is carried out on the original data (rather than on the residuals from fitting additive terms). We thus find that for the pentosans data DF = 20.95 for the first term. Similarly, we can find DF values for the second, third, . . . , terms. Table 9.7 shows the results. It is clear from this table that the mean square for the first multiplicative component of interaction is larger than that of subsequent terms by an order of magnitude; it cannot therefore represent random error. The subsequent terms can, however, be pooled by dividing the sum of their sums of squares by the sum of their degrees of freedom. The pooled value is an estimate of the variance of the error term. We thus obtain: $\sigma = \sqrt{0.0262} = 0.16$.

In Chapter 10 we will approach the problem of interlaboratory testing in a more fundamental way for the pentosans data. We will see that many characteristic features of the data can be revealed only by an analysis that is more particularly geared to the subject of interlaboratory testing.

9.10 SUMMARY

In this chapter we have presented a very general technique for analyzing two-way tables. We have also discussed the relation between this technique and other methods described in previous chapters.

REFERENCES

Gabriel, K. R. (1971). The Biplot Graphical Display of Matrices with Application to Principal Component Analysis. Biometrika *58*, 453–67.

Gabriel, K. R. and Dan Bradu (1978). The Biplot as a Diagnostic Tool for Models of Two-Way Tables. Technometrics *20*, 47–68.

Mandel, John (1971). A New Analysis of Variance Model for Non-Additive Data. Technometrics *13*, 1–18.

10

Interlaboratory Studies

10.1 GENERAL CONSIDERATIONS

The usefulness of a measurement process depends largely on its stability, both in time and in space. Control charts, as we will see, are an excellent technique for studying their stability in time. They address the question. "How much does the process vary (or shift, or cycle, etc.) with time?" Both the question and the answer provided by the control charts are, however, generally limited to a single laboratory. This is often satisfactory, since in many cases, the measuring is done for the purpose of insuring the stability of a *manufacturing process,* the one taking place in the plant with which the laboratory is connected. There are situations, however, where more is demanded of the measuring process. This occurs whenever the measurement is made for the purpose of testing a product for conformance with established specifications, or for the purpose of ascertaining whether two laboratories, for example those of the vendor and of the buyer, have obtained compatible test results when measuring a quality characteristic of the product that is sold and purchased. We can refer to these matters as dealing with the stability of the measuring process in space rather than in time.

The procedure that has been developed for the purposes just discussed is called *interlaboratory testing*. It consists in sending materials to a number of laboratories, generally laboratories that participate voluntarily and without remuneration in the test, and collecting, analyzing, and interpreting the results obtained by these laboratories.

A typical example of the results of an interlaboratory study is the set shown in Table 1.2. Seven laboratories participated in this study. Each laboratory received portions of each of nine materials, and performed the test in triplicate on each material.

In general, p laboratories will test q materials, and perform n replicate tests on each material. The materials should be chosen so as to cover the entire range of interest, as shown by the example. Each material should be made as homogeneous as possible prior to distribution to the laboratories. A replicate (test result) consists in carrying out *all* the steps of the measuring process (not just part of it). Thus if the test is a chemical analysis consisting of a *wet chemistry* part, followed by an instrumental, physical measurement such as the reading of an absorbance at a specified wavelength of the spectrum, each replicate should include both of these phases, not just the reading of the absorbance.

The laboratories should all be reputable, competent, well-equipped, and acquainted with the measurement process under study. Occasionally this last condition is not satisfied. In this case, a preliminary *familiarization* experiment should precede the interlaboratory study proper.

It is imperative that a detailed description of the measurement technique be available in written form, including all its steps, and that identical copies of it be sent to all participating laboratories, so as to avoid as much as possible, misinterpretations or different versions of the process. It is also important to prepare a *protocol* of the interlaboratory study, consisting of all pertinent instructions, in addition to the description of the test procedure, to the participating laboratories.

Matters such as the number n of replicate test results, the number of significant figures to be reported, the *repeatability conditions* (replicates run on same day, or on different days, with or without recalibration of the instrument, etc.) are part of the protocol.

A central laboratory or committee should be responsible for the conduct of the entire operation, and for finding a qualified individual for the analysis and interpretation of the data.

An important point is the need to examine the data promptly, so as to be able to get in touch with the participating laboratories as soon as possible after their testing activities, to clarify (and possibly correct) any anomalies observed in the data. The laboratories themselves should not throw out any *outlying* observation, unless something went wrong during the measurement process, in which case they should repeat the test result(s) regardless of whether they *look right*. It is useful to prepare data-reporting sheets, identical for all laboratories, including an encouragement, and appropriate space on the sheets, for pertinent observations by the technicians who perform the tests. Finally, each laboratory should designate a person who is responsible for the operations carried out in it in regard to the interlaboratory study, and with whom the central committee will communicate personally, or by mail or telephone.

10.2 PRECISION AND ACCURACY

An ideal measuring process, when applied to an unending series of samples from the same homogeneous material, would produce the same value, namely the "correct" value, every time. Such a process does not exist in reality. In such an unlimited series of measurements, a real measuring process would generate a statistical population of measurements, as we discussed in the beginning of this book. The smaller the spread of this population (as measured, for example, by its standard deviation), the more *precise* will be the measuring process. If, moreover, the mean of the generated population, say μ, coincides with the "true value" of the measured characteristic, the process will be "unbiased" or, as some would say, *accurate*. If the mean μ is not identical with the true value, say T, then the process has a *bias* equal to $\mu - T$.

The weakness in the above discussion lies in its assumption that the hypothetical repetition of the measurement process would indeed produce a single, well-defined, statistical population. We will see how this assumption can be checked, through the aid of control chart techniques, in terms of stability in time. If, on the other hand, we imagine the process to be carried out in different *locations,* e.g., different laboratories, it is not unreasonable to consider each location as generating its own, individual statistical population. Then, the term precision has to be redefined: it is now characterized, not only by the smallness of the standard deviation of each generated population, but also by the closeness to each other, of the means of the populations corresponding to the various locations (laboratories). Thus we consider two components of precision: within-laboratory precision and between laboratory precision. Since precision refers to the *smallness* of spread, it is easier to speak of imprecision, and to consider two components of imprecision, or, more simply, of variability.

One could, of course, consider more than two components, including such sources of variability as *days within laboratories,* operators, instruments, and others. It is unwise to do so. Indeed, days within laboratories refers to a time-related variability and is best studied by control chart techniques within each laboratory. The effect of operators, instruments, etc., is likely to vary considerably from laboratory to laboratory, and cannot be measured globally in an interlaboratory study. The best course of action is, therefore, to limit oneself to the two components of within- and between-laboratory variability, known respectively as *repeatability* and *reproducibility*. As mentioned earlier, the "repeatability conditions" should be specifically spelled out in the protocol, to obtain a reasonably reproducible *within-laboratory variability* among laboratories.

It has been found that for most measuring processes, repeatability and reproducibility, when quantified (e.g., as standard deviations) are functions of the *level* of the measured quantity. This is the main reason for running an interlaboratory study of a test method at a variety of levels, each represented by a specific *material*. Thus, in our example, the levels varied from a value of pentosans content of less than 1 percent to one of about 16 percent.

10.3 ANALYSIS OF THE DATA

It is now clear that the main objective of the analysis is to measure the within- and between-laboratory components of variability, and to do this at each level separately. This leads us at once to what we will call the "level-by-level analysis." It consists of a within-between analysis of variance carried out separately for each of the q levels included in the study, following the method we have discussed earlier in this book. It is very advantageous to reorder the materials in increasing order of magnitude of the measured characteristic, in reporting the results of the analyses, as well as in subsequent graphs. We have done this for the pentosans data in Table 1.2. The original order of the materials was different. The results of this analysis are shown in Table 10.1 and plotted in Fig. 10.1. The symbol $s_r(j)$ is the square root of the pooled value of within-laboratory variance over all laboratories, for level j. The symbol $s_R(j)$ is equal to $\sqrt{s_r(j)^2 + s_L(j)^2}$ where $s^2_L(j)$ is the component of variance between laboratories at level j. The only remaining task is to fit appropriate straight lines or curves to the plotted points, which will permit interpolation at any level not specifically covered in the interlaboratory study. We will do this in Section 10.8.

The analysis we have just carried out is subject to a number of criticisms, both of a theoretical and practical nature. We will therefore

Table 10.1 Level by Level Analysis of Pentosans Data (Table 1.2)

Level	Average	$s_r(j)$	$s_R(j)$
A	0.405	0.015	0.114
B	0.887	0.034	0.049
C	1.128	0.143	0.196
D	1.269	0.037	0.074
E	1.981	0.040	0.063
F	4.181	0.033	0.209
G	5.184	0.133	0.243
H	10.401	0.194	0.585
I	16.361	0.216	1.104

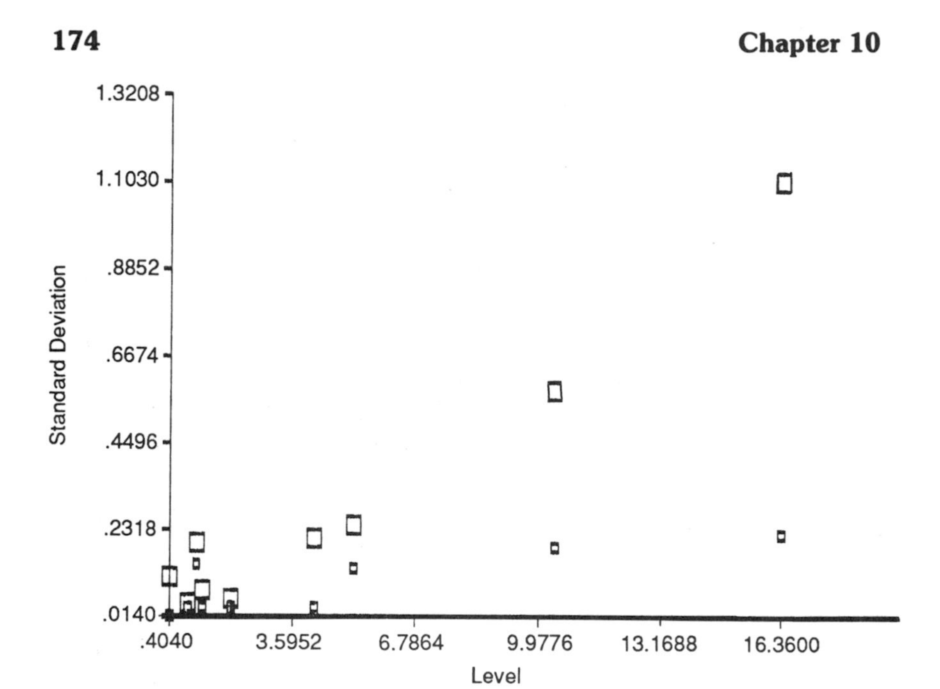

Figure 10.1 Pentosans in pulp. Graph of $s_r(j)$ and $s_R(j)$, repeatability and reproducibility versus level.

present additional analyses as well as a different approach. At present, we discuss some shortcomings of the within-between, level-by-level analysis.

10.4 CRITIQUE OF THE LEVEL-BY-LEVEL ANALYSIS

From a theoretical point of view, the level-by-level analysis is flawed in the following way.

As observed by R.A. Fisher (1935), there is a fundamental difference between deductive and inductive inferences. In the former, as exemplified by the theorems of Euclidean geometry, conclusions can be drawn in an absolute way from a set of premises, and additional premises have no effect on these conclusions. Thus, given a triangle in a plane, one can prove, from the postulates of Euclidean geometry, that the sum of its angles is equal to 180 degrees. Any additional information, such as for example, the lengths of the sides of the triangle, cannot change the conclusion.

In inductive inference, as exemplified by statistical data analysis, the situation is quite different. From a given set of data, together with given assumptions, one can draw inferences. Thus we can *estimate* a population standard deviation, by calculating the standard deviation of a random sample from this population, say a sample of size 10. If, however, we take a second sample, say of size 5, from the same population, the proper estimate of the population standard deviation is now a new estimate, calculated from the combined sample of size 15. Generally, this will yield a different numerical value.

It follows that in inductive inference situations, any inference should be based on *all* the information available, not just part of it.

Applying these considerations to the level-by-level analysis of interlaboratory data, we see that this analysis ignores the fact that the *same* laboratories measured *all* materials. Even Table 10.1 and Fig. 10.1, in which the conclusions drawn from all levels are finally brought together, do not take into account that a single set of laboratories was involved in the entire study. Had we assigned the testing of each material to a different set of p laboratories, thus involving $p \times q$ laboratories, rather than p, the analysis would not reflect this fact.

Turning now to more concrete considerations, we have not examined the *individual* results in our analysis, but rather applied a global, noncritical analysis to them.

Thus, it might have turned out that at any one level, 8 of the 9 laboratories gave results very close to each other, while the one remaining laboratory obtained very different results. Is the standard deviation of the sample of nine laboratory averages then still a valid one? Similarly, the three test results in some given cell may differ from each other by much more than the replicates in other cells at the same level. Yet, the within-laboratory variance is a *pooled* value over all laboratories. How valid is such a pooled value under such circumstances?

It is clear from the above considerations that the level-by-level analysis, in order to be valid, must be supplemented by additional procedures, geared to a more detailed scrutiny of the entire body of data, taking into account all relevant features of the design of the interlaboratory study.

10.5 OUTLIERS

Our first task, in supplementing the naive level-by-level analysis, is to scrutinize the data for outliers. Generally speaking, an outlier is a single test result, or a group of test results, that seem to be drastically different from the bulk of the data. This could be due to *blunders,* such as writing 61.2 instead of 16.2, or to a set of circumstances, such as a temporary power failure, that vitiated the analysis, or to a mislabeling of a vial or a test specimen. Provided the examination of the data is carried out promptly after the measuring activities, it is generally possibly to "catch" the reasons for such blunders, especially if the precautions listed in Section 10.1 have been taken. The outliers that cause real problems are those resulting from more subtle causes, and are not so drastic as to be obviously classified as blunders.

We will discuss two types of outliers and see that either type may lead to a third category.

10.6 THE *k*-STATISTIC

The first type of outlier we discuss consists in test results within a single cell that show considerably poorer (or sometimes, better) agreement with each other than other cells for the same material. To detect such occurrences, we need a descriptive statistic. We choose the statistic, which we call *k,* defined as follows:

$$k = \frac{s_{ij}}{(s_r)_j} \tag{10.1}$$

Here s_{ij} is the standard deviation among replicates within cell (i,j), and $(s_r)_j$ is the pooled value for replication standard deviation for material (level) j. Since s_{ij} was included in the pooling operation leading to $(s_r)_j$, the numerator and denominator of k are not statistically independent, but this nonindependence can easily be dealt with in deriving the frequency distribution of k.

Table 10.2 Pentosans Data (Table 1.2), k-Values

Lab	A	B	C	D	E	F	G	H	I
1	1.926	2.126	2.606	2.619	2.316	0.710	2.474	0.344	1.526
2	0.000	0.170	0.000	0.154	0.668	0.178	0.000	0.717	0.209
3	0.000	0.170	0.081	0.000	0.636	0.888	0.217	0.477	0.232
4	1.019	0.340	0.081	0.000	0.146	0.355	0.000	1.211	0.608
5	0.000	0.900	0.000	0.000	0.292	1.627	0.174	0.542	0.640
6	1.019	0.681	0.040	0.154	0.386	1.517	0.230	0.149	0.844
7	1.102	1.017	0.443	0.308	0.729	0.774	0.867	2.087	1.756

Critical values for k, at three levels of significance (1%, 0.5%, and 0.1%) are given in Table A1 of the Appendix. It is assumed in this table that the within-laboratory errors are normally distributed. A k-value that shows significance at the 0.5% level will result in "flagging" the cell to which it corresponds. The flagging is merely a guide, useful in judging the seriousness of the discrepancy of the cell in question, and *not* a rejection device. Since to each cell there corresponds a k-value, it is useful to tabulate all $p \times q$ k-values, as shown in Table 10.2, and to prepare two plots of the k-values. These are shown in Figs. 10.2 and 10.3. In the first, the k-values are grouped by levels, and given in the order of the laboratories within each group. Thus, in this plot, the groups are independent, but the bars within each group are mutually dependent. A very large k-value in any one such group is compensated by small k-values for the remaining laboratories within the group. (It can easily be seen that the sum of the squares of k within each group is equal to the number of laboratories, p). The

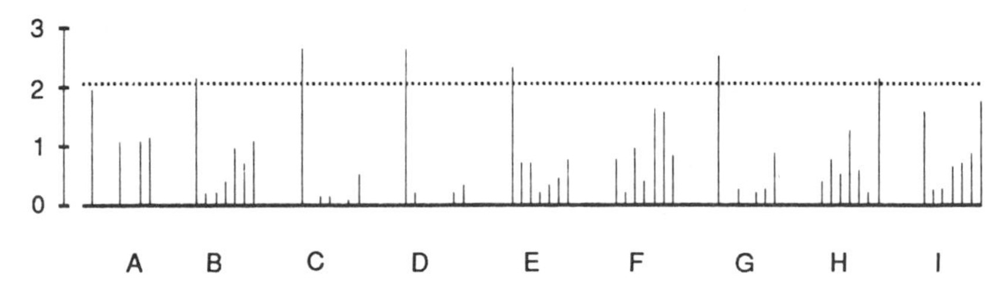

Figure 10.2 Pentosans in pulp; graph of k-values by materials.

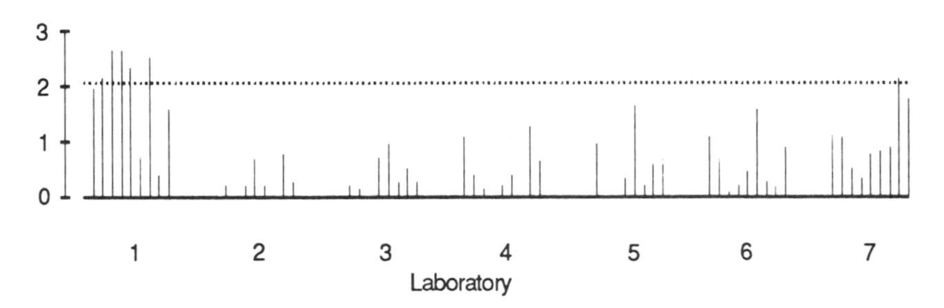

Figure 10.3 Pentosans in pulp; graph of k-values by laboratories.

graph provides therefore additional means of flagging an outlying k-value.

Of more interest to us is Fig. 10.3, where the same k-values have been rearranged in groups of laboratories, the k-values within each group being plotted in the order of levels. Now the bars within each group are statistically independent and therefore provide valid cumulative evidence for the behavior of the laboratories. For our data, Fig. 10.3 provides convincing evidence for the conclusion that the within-laboratory variability varies systematically, and rather considerably from laboratory to laboratory. This conclusion, incidentally, at once raises questions as to the validity of the one-way analysis of variance, which is based on the assumption of *homoscedasticity* (equality of variance) for the within-laboratory component.

From a practical point of view, Fig. 10.3 should lead to action on the part of the committee running the study. The reasons for the discrepancies should be investigated, and this may well lead to a revision of the test method description, aimed at getting better agreement for replication error (within-laboratory variability) between laboratories. Such action is far superior to the automatic *rejection* of cells (or laboratories) on the basis of tests of significance, a procedure that simply ignores the observed difficulties instead of using them for the purpose of improving the test method.

We also see that our examination of k-values has led, from a consideration of individual discrepant cells, to one of an entire discrepant laboratory. This is the third type of outliers we mentioned earlier. In the case of our data, we are especially struck by the behavior of laboratories 1 and 7.

Since we are dealing with data obtained many years ago, all we can do here is to use the analysis for a discussion of actions that should have been taken at the time the study was run, and proceed with our analysis.

10.7 THE *h*-STATISTIC

The second type of outlier we consider relates, not to the variability within cells, but rather to the location of an entire cell in relation to the other cells at the same level. This is indicated by a *cell average* that shows considerable discrepancy from the remaining cell averages at the same level. The statistic we use is the *h-statistic* defined by

$$h = \frac{y_{ij} - x_j}{(s_L)_j} \tag{10.2}$$

Here, y_{ij} is the average of all test results in cell (i,j); x_j is the grand average of all test results at level j (or, in other words, x_j is the average of all y_{ij} for level j); and $(s_L)_j$ is the standard deviation among the y_{ij} for level j.

The numerator of h, $y_{jj} - x_j$, is simply the *deviation* of cell (i,j) from the "consensus-value" x_j at the same level. The denominator $(s_L)_j$ is a *standardizing* device, motivated by the fact that the deviations generally vary considerably in magnitude from level to level.

Unlike k, which is by definition, a positive quantity (unless it is zero), h can be positive or negative. A positive value of h indicates a cell average that lies *above* the consensus value; and a negative h, a cell average that lies *below* the consensus value.

As we did for the k-statistic, we prepare a table of all $p \times q$ h-values (see Table 10.3) and plot the values in two ways, by groups of levels, and by groups of laboratories. These plots are shown in Figs. 10.4 and 10.5.

The same considerations of dependence and independence apply here as for the k-values. We note that the sum of squares of all h-values at a given level is equal to $p - 1$.

As was the case for the k-statistic, the plot of real interest is the second one, (Fig. 10.5), in which the groups represent laboratories. This graph shows not only which individual cells are discrepant, but also the behavior pattern of the different laboratories in terms of *lo-*

Table 10.3 Pentosans Data (Table 1.2), *h*-Values

Lab	A	B	C	D	E	F	G	H	I
1	0.459	0.318	2.049	0.564	−1.505	−0.168	1.730	0.632	0.357
2	0.046	−1.311	−0.051	−0.226	−0.390	−0.377	0.349	−0.748	−0.255
3	0.931	0.888	−0.072	1.205	1.346	−0.184	−0.035	−0.505	−0.322
4	−0.190	1.458	0.055	0.317	1.160	0.122	0.072	0.573	0.375
5	0.754	−0.985	−0.942	−0.571	−0.514	1.973	−0.910	−0.043	−0.692
6	0.076	0.155	−0.094	0.564	0.230	−1.375	−1.417	−1.446	−1.304
7	−2.076	−0.521	−0.945	−1.853	−0.328	0.009	0.211	1.538	1.840

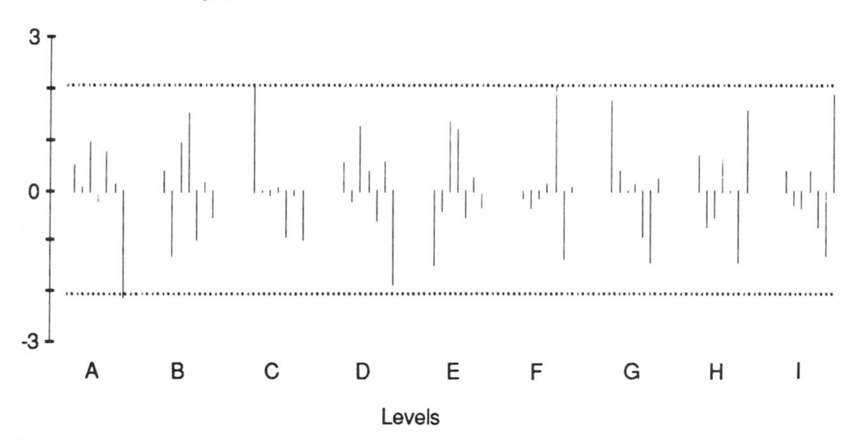

Figure 10.4 Pentosans in pulp; graph of h-values by materials.

cation. (The k-statistic related to the behavior pattern of the laboratories in terms of within-cell variability). Some laboratories obtain test results that are systematically low (i.e., nearly predominantly *below* the consensus value) while others obtain systematically *high* test results. Still others may go from low to high, or the reverse, as the level increases. Here again, the committee should take immediate action, where action is required. It is generally impossible to eliminate completely the systematic variability among laboratories, but laboratories that are drasti-

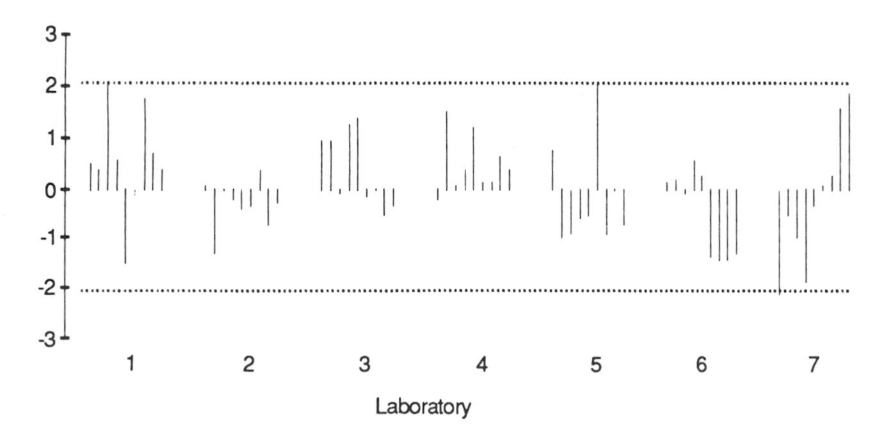

Figure 10.5 Pentosans in pulp; graph of h-values by laboratories.

cally different from all others should be examined for possible non-compliance with either the test method description itself or with other parts of the protocol. Thus, here as in the case of the k-statistic, the purpose is not *rejection* on the basis of significance tests, but rather a detection of systematic patterns, possibly followed by action aimed at improving the test method or the interlaboratory study protocol. As mentioned above, such action is only possible at the time the study is made, and our analysis of the pentosans data is therefore limited to pointing out where action should have been taken at the time the study was made. The real purpose of our discussion is to alert the reader to what should be done when a new round robin is conducted, apart from merely calculating and recording precision parameters relating to repeatability and reproducibility. We also note that the h- and k-statistics go a long way towards supplementing the simple within-between analysis of variance with a detailed examination of the data. The h- and k-statistics extract much additional information from the data. Thus, they are far more than "tests for outliers."

10.8 REPEATABILITY AND REPRODUCIBILITY AS FUNCTIONS OF LEVEL

In Section 10.3 we discussed the dependence of s_r and s_R on the level. Figure 10.1 shows this dependence for the data of Table 10.1. Assuming that no data have been eliminated as a result of the analysis (h and k graphs), Fig. 10.1 can lead to numerical equations for the expression of the dependence of s_r and s_R on the level x. Note first that the information contained in Fig. 10.1 is *not* contained in either h or k, since both of these are obtained from level-standardized data. Thus Fig. 10.1 is new information. Secondly, when fitting any curve to points such as those shown in Fig. 10.1, it should be remembered that the statistical population of a standard deviation has a variance that depends on the mean of this population.

Let s represent a standard deviation based on a sample of size $\nu + 1$ from a normal population and let σ be the corresponding true standard deviation. Then it can be shown that, on the assumption of normally-distributed errors, the quantity $\nu s^2/\sigma^2$ follows a well-known distribution known as *chi-square with* ν *degrees of freedom* and denoted χ^2_ν. Thus

$$\frac{\nu s^2}{\sigma^2} = \chi^2_\nu \tag{10.3}$$

Now it can also be shown that the mean of χ^2_ν is ν, and that its variance is 2ν. From Eq. (10.3) we obtain

$$s^2 = \frac{\chi^2_\nu}{\nu}\,\sigma^2 \tag{10.4}$$

Therefore we have:

$$\text{Mean of } s^2 = \frac{\nu}{\nu}\,\sigma^2 = \sigma^2$$

$$\text{Variance of } s^2 = \frac{2\nu}{\nu^2}\,\sigma^4 = \frac{2\sigma^4}{\nu}$$

Consequently, the standard deviation of s^2 is equal to

$$\text{Standard deviation of } s^2 = \sigma^2\sqrt{\frac{2}{\nu}} \tag{10.5}$$

However, we need the standard deviation of s, not of s^2. We have,

$$s^2 - \sigma^2 = (s + \sigma)(s - \sigma)$$

Therefore

$$s - \sigma = \frac{s^2 - \sigma^2}{s + \sigma} \cong \frac{s^2 - \sigma^2}{2\sigma} \tag{10.6}$$

The standard deviation of s is the same as that of $s - \sigma$, which according to Eq. (10.5), is approximately that of $s^2 - \sigma^2$, divided by 2σ. We then have:

$$\text{Standard deviation of } s^2 = \sigma^2\sqrt{\frac{2}{\nu}}$$

and

$$\text{Standard deviation of } s = \sigma^2\,\frac{\sqrt{2/\nu}}{2\sigma} = \frac{\sigma}{\sqrt{2\nu}} \tag{10.7}$$

Table 10.4 Weighted Linear Regression of $(s_r)_j$ on x_j for Pentosans Data (Table 1.2)

x	s	$\hat{s}_1$ [a]	w_1 [b]	$\hat{s}_2$ [c]	w_2 [d]	$\hat{s}_3$ [e]
.405	.015	.0435	528	.0417	575	.0415
.887	.034	.0492	412	.0477	440	.0475
1.128	.143	.0521	368	.0507	389	.0506
1.269	.037	.0538	346	.0524	363	.0524
1.981	.040	.0622	258	.0613	266	.0613
4.181	.033	.0884	128	.0888	127	.0889
5.184	.133	.1003	99.4	.1012	97.5	.1015
10.401	.194	.1623	38.0	.1663	36.2	.1670
16.361	.216	.2331	18.4	.2406	17.3	.2416
Intercept		.0387		.0366		.0364
Slope		.01188		.01246		.01254

[a] Unweighted regression.
[b] $w_1 = 1/\hat{s}_1^2$.
[c] First weighted regression.
[d] $w_2 = 1/\hat{s}_2^2$
[e] Second weighted regression.

Now, in Fig. 10.1, the standard deviations cover a fairly wide range, from 0.01 to 1.1. Therefore, their uncertainty is not constant, as shown by Eq. (10.7). The weight of each s is, approximately $2\nu/\sigma^2$, hence proportional to $1/\sigma^2$. A simple way to fit a curve to the points is an iterative procedure.

Let $s = f(x)$ be the curve to be fitted. First, fit it without weights. This will yield a function $\hat{s} = \hat{f}(x)$. Then, repeat the fit, using weights equal to $1/\hat{s}^2$. If necessary, repeat the entire process. The procedure is quite simple with a computer. It is illustrated in Table 10.4, which shows that the number of iterations required is, in this case, quite small. This is generally the case.

10.9 THE ROW-LINEAR MODEL FOR INTERLABORATORY TEST DATA

While the h- and k-plots are effective means for extracting more detailed information from the data, and the plot of s_r and s_R versus the level x provides valuable additional information, there are features of the data that are not revealed by these graphs, and others that are only marginally apparent in them. Thus, for the pentosans data, the h-plot (Fig. 10.5) seems to show a systematic change in h for laboratory 7, from large and negative at low pentosans levels, to large and positive at high levels. Such trends can be studied more effectively by a different approach to the analysis of interlaboratory data, an approach based on the row-linear model, with which we are already familiar. The row-linear model is not concerned with within-cell replication, a matter that is satisfactorily dealt with by the k-statistic. We therefore start with a table of cell averages, shown in Table 10.5 for the pentosans data. At the bottom of the table we have added a row, labeled x_j, consisting of the column averages, i.e., the *consensus values* at all levels.

If we consider a particular laboratory, say laboratory 1, we can obtain useful information by plotting the results obtained by this laboratory (in terms of cell averages) against the corresponding consensus values x_j. Such a plot is shown in Fig. 10.6, and suggests that a straight line would be a reasonable fit for the nine points.

The usual equation for a straight line fit is (see Section 5.2):

$$y_j = \hat{\alpha} + \hat{\beta}\, x_j + d_j \tag{10.8}$$

where the estimates $\hat{\alpha}$ and $\hat{\beta}$ for α and β are given by Eqs. (5.6b) and (5.6a), and d is the residual.

The estimates $\hat{\alpha}$ and $\hat{\beta}$ can be shown to be correlated. It is therefore useful to write the fitted straight line equation in a slightly different form, as follows (see Eq. 5.9):

$$y_j = \bar{y} + \hat{\beta}\,(x_j - \bar{x}) + d_j \tag{10.9}$$

Comparing Eqs. (10.8) and (10.9), we see that

$$\hat{\alpha} = \bar{y} - \hat{\beta}\,\bar{x}$$

where $\hat{\beta}$ is given by Eq (5.6a). The fitted line is given by

$$\hat{y}_j = \hat{y} + \hat{\beta}\,(x_j - \bar{x}) \tag{10.10}$$

Table 10.5 Pentosans Data (Table 1.2), Table of Cell Averages

Lab	A	B	C	D	E	F	G	H	I
1	0.457	0.900	1.450	1.307	1.900	4.147	5.560	10.757	16.750
2	0.410	0.833	1.120	1.253	1.960	4.103	5.260	9.980	16.083
3	0.510	0.923	1.117	1.350	2.053	4.143	5.177	10.117	16.010
4	0.383	0.947	1.137	1.290	2.043	4.207	5.200	10.723	16.770
5	0.490	0.847	0.980	1.230	1.953	4.590	4.987	10.377	15.607
6	0.413	0.893	1.113	1.307	1.993	3.897	4.877	9.587	14.940
7	0.170	0.866	0.980	1.143	1.963	4.183	5.230	11.267	18.367
x_j	0.405	0.887	1.128	1.269	1.981	4.181	5.184	10.401	16.361

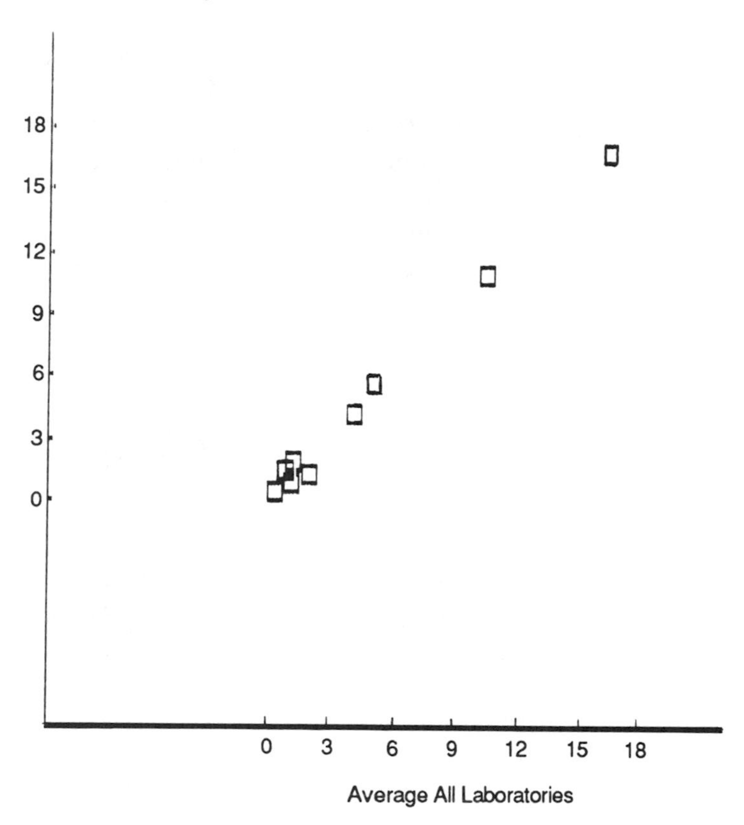

Figure 10.6 Pentosans in pulp, lab 1 versus average all laboratories.

When $x_j = \bar{x}$, $\bar{y}_j$ becomes $\bar{y}$, so that the point $(\bar{x},\bar{y})$ is a point on the fitted line. This point is called the *centroid* of the set of points that are being fitted.

Returning now to our pentosans cell averages, we can fit a line separately for each laboratory, by plotting, for each laboratory, the cell averages obtained by that laboratory against the corresponding x_j (which are therefore the common abscissa values for all p fits). The parameters for each fit are $\bar{y}$ and $\hat{\beta}$, and they will in general, differ from laboratory to laboratory. Thus, the general equation for the model we are describing is

$$y_{ij} = \bar{y}_i + \hat{\beta}_i (x_j - \bar{x}) + d_{ij} \qquad (10.11)$$

This is, with a slightly different notation, the model we have called *row-linear* in Chapter 7. We have called this model the row-linear model because all rows in the table (of cell averages) are considered to be linearly related to the x_j (column averages). The departures of the data from this model are the residuals d_{ij}; therefore, a careful examination of the residuals is essential for judging whether the row-linear model is indeed a satisfactory fit. Table 10.6 is a display of the residuals and Fig. 10.7 is a plot of standardized residuals by laboratories.

10.10 THE PARAMETERS OF THE ROW-LINEAR MODEL

In accordance with Eq. (10.11), we list in Table 10.7 the values of $\bar{y}$ and $\hat{\beta}$ for all seven laboratories. It is evident that there is a relationship, and Fig. 10.7 shows that it can be approximated by a straight line. But according to what we learned in Section 7.5, this means that the seven lines corresponding to the seven laboratories *concur* approximately in a point whose abscissa and ordinate are both equal to $\bar{x} - 1/\gamma$, where γ is the slope of the line of $\hat{\beta}$ versus $\bar{y}$. The straight line relation between $\bar{y}$ and $\hat{\beta}$ is by no means a general situation. It happens to be the case for our data. However, when it occurs it always indicates concurrence.

Denoting by y_0 the abscissa and the ordinate of the point of concurrence, we have

$$y_0 = \bar{x} - \frac{1}{\gamma}$$

where γ is the slope of the line of $\hat{\beta}$ versus $\bar{y}$, and the model for our data becomes:

$$\hat{y}_{ij} - y_0 = \frac{(\bar{y}_j - y_0)\,(x_j - y_0)}{\bar{x} - y_0}$$

For our data, we find

$$\gamma = 0.3495 \qquad y_0 = 1.78$$

Table 10.6 Residuals From Row-Linear Fit Pentosans Data (Table 1.2)

Lab	A	B	C	D	E	F	G	H	I
1	−0.0070	−0.0573	0.2460	−0.0412	−0.1770	−1.1827	0.2041	0.0621	−0.0459
2	−0.0024	−0.0504	0.0007	−0.0033	0.0071	−0.0004	0.1761	−0.2027	0.0753
3	0.0271	−0.0279	−0.0687	0.0282	0.0397	−0.0073	0.0521	−0.0741	0.0311
4	−0.0086	0.0596	0.0021	0.0112	0.0331	−0.0627	−0.0990	0.0684	−0.0041
5	0.0083	−0.0992	−0.1978	−0.0831	−0.0455	0.4731	−0.0955	0.2732	−0.2336
6	−0.0734	−0.0312	−0.0300	0.0357	0.0756	−0.0188	0.0507	0.0245	−0.0332
7	0.0559	0.2063	0.0478	0.0525	0.0670	−0.2011	−0.2884	−0.1504	0.2104

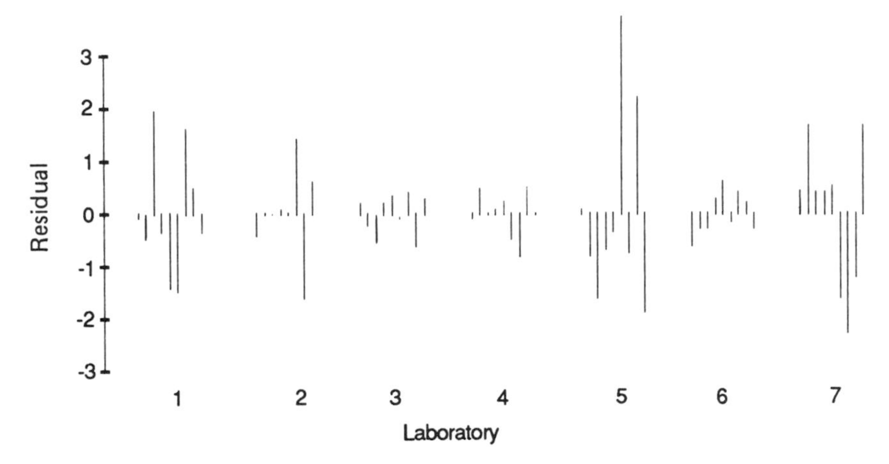

Figure 10.7 Pentosans in pulp; graph of studentized residuals by laboratories. Residuals from linear model fit.

10.11 ANALYSIS OF VARIANCE FOR THE CONCURRENT MODEL

It can be shown that the analysis of variance can be further extended to include the breakdown of the sum of squares for slopes (row-linear model) into a sum of squares for "concurrence" and a sum of squares for "nonconcurrence."

Table 10.7 Pentosans Data (Table 1.2), Linear Model Analysis

Lab	$\bar{y}$	$\hat{\beta}$
6	4.3346	0.9079
2	4.5559	0.9774
5	4.5622	0.9625
3	4.6000	0.9712
4	4.7444	1.0267
1	4.8030	1.0236
7	4.9076	1.1307

Table 10.8 Pentosans Data (Table 1.2) Analysis of Variance

Source	DF	SS	MS
Rows	6	1.947392	.3245653
Columns	8	1637.082	204.6352
$R \times C$	48	7.904026	.1646672
Slopes	6	6.919914	1.153319
Conc.	1	6.181299	6.181299
Nonconc.	5	.7386151	.147723
Resid.	42	.9841189	.0234314

For a row-linear model, the sum of squares for "slopes" (row linearity) is, as we have seen, equal to

$$SS_{\text{slopes}} = \sum_i (B_i - 1)^2 \sum_j (C_j - M)^2$$

Now, we can write

$$B_i - 1 = \gamma(R_i - M) + \Delta_i$$

This corresponds to the regression of B_i on R_i with slope γ and residual Δ_i. Exact concurrence requires that Δ_i be very small with respect to the term $\gamma(R_i - M)$. We have in all cases:

$$\sum_i (B_i - 1)^2 = \gamma^2 \sum_i (R_i - M)^2 + \sum_i \Delta_i^2$$

Therefore

$$SS_{\text{slopes}} = \gamma^2 \sum_i (R_i - M)^2 \sum_j (C_j - M)^2 + \sum_i \Delta_i^2 \sum_j (C_j - M)^2$$

It can be shown that the degrees of freedom corresponding to the two terms $\gamma^2\Sigma(R_i - M)^2\ \Sigma(C_j - M)^2$ and $\Sigma\Delta^2{}_i\Sigma(C_j - M)^2$ are 1 and $p - 2$, respectively, where p = number of laboratories. Table 10.8 is an analysis of variance of the pentosans data incorporating this further partition. We see that the *slopes* sum of squares is almost entirely due to concurrence. The mean square for Concurrence is very large when compared to that for Nonconcurrence. Nevertheless, the latter is still larger than that for *residual,* which shows that while the data are essentially concurrent, there remain some unexplained sources of variability. The analysis of variance table is useful, but does not show the behavior of the data in detail. It must always be supplemented by an additional investigation of the data.

10.12 CONCLUSION

We have analyzed the pentosans data (Table 1.2) in three different ways: as a row-linear model, which turned out to be concurrent; as an additive plus multiplicative model, in which one multiplicative term appeared to be sufficient; and by a method especially geared to interlaboratory data, with *h*- and *k*-graphs. Each method showed some details that were not apparent in the other analyses. Certainly the analysis of variance tables, which are generally considered "exact" statistical methods, were shown to be deficient. This is so because they are global methods, exhibiting sums of squares rather than the behavior of individual data. An analysis of variance table by itself, is always insufficient as a method of data analysis, unless supplemented by additional techniques. The *k*-graph by laboratories showed very clearly that laboratory 1, and to some extent laboratory 7, had poorer repeatability of results than the remaining five laboratories. This in itself invalidates to a certain extent a pooled repeatability standard deviation. The *h*-graph exhibited a systematic effect of laboratories. It also showed other peculiarities of the data, for example the definite trend for laboratory 7, from values below average to values well above the average. The graph of s_r and s_R versus material average is consistent with a concurrent model by exhibiting a linear trend of s_R with a positive slope. Indeed, in a concurrent model, with the point of concurrence to the left, there must be a gradual widening of the bundle of straight lines from left to right, resulting in an increasing standard deviation between the lines (laboratories).

We conclude that it is often useful to look at data in a number of different ways, using different statistical techniques. It is also apparent that real data seldom follow a single theoretical pattern. Mathematical models may be aesthetically very satisfying; they rarely represent real data sets.

REFERENCE

Fischer, R.A. (1935). The logic of inductive inference. J. Royal Statistical Society, *98*, 39–54.

11

Control Charts

11.1 INTRODUCTION

The control chart is an invention of Walter Shewhart (1931). Its aim is to control the quality of a production process. One of the reasons for including it in this book is to show that it is also a very effective tool for the control of a measuring process.

Shewhart conceived of two control charts, one for averages and one for ranges. We will follow him quite closely, except that we replace the range chart by one of standard deviations.

Essentially, a control chart deals with data classified according to groups. Generally, the groups are time intervals, and the study of these time-groups is the primary function of control charts. However, we will show that control charts are also an effective tool to the data analyst, even in situations in which the groups do not involve time. Therefore, we will present two applications of control charts, one involving time, and the second involving an entirely different criterion of classification. Following this presentation, we will discuss control charts for discrete data (counts).

11.2 AN EXAMPLE INVOLVING TIME INTERVALS

Table 11.1 contains a set of artificial data in terms of averages and standard deviations. In each of 30 time periods constituting the *baseline,* 4 replicate measurements were made. The values in Table 11.1 are the averages and standard deviations of each set of 4. The data are plotted in Fig. 11.1, for averages ($\bar{x}$) (top) and for standard deviations (s) (bottom). Each of these figures contains 10 additional points listed in Table 11.2, each based on 4 measurements, obtained after the base period of the first 30 points. On each chart there are 3 horizontal lines: the "central line" and two "2-*sigma control lines.* The central line for $\bar{x}$ is drawn at the grand average, $\bar{\bar{x}}$ of the 30 $\bar{x}$ values. The central line for the s-chart is drawn at the value $\bar{s}$, the *average* of all 30 s-values. Here, control chart work differs from what we have stated earlier for the pooling of standard deviations. The control lines for the s-chart are obtained by muliplying $\bar{s}$ by modifications of the factors B_7 and B_8, respectively. The values for B_7 and B_8 are recorded in Table A2 of the Appendix. The probability of a point falling outside the control lines by pure chance is of the order of five percent.

The control lines for the average chart are derived from the s-chart. They are obtained by multiplying $\bar{s}$ by a modification of respectively A_3 and $(-A_3)$, where A_3 is also listed in Table A2. The modified multiplier A_3 is such that the probability of an $\bar{x}$ point falling outside the control band by pure chance is of the order of five percent. This statement requires clarification. It means that if the variability of the $\bar{x}$ points is due to *no other cause* than that reflected in the standard deviations, the $\bar{x}$ values should fall inside the control band with probability of 95 percent. It is of course possible that other sources of variability affect $\bar{x}$. In such cases, the limits should be obtained by calculating the standard deviation s_x of all the $\bar{x}$ points in the base period, and multiplying it by $(+\ 2)$ and $(-\ 2)$. In its original form (control lines based on the s-chart), the $\bar{x}$-chart is essentially a test as to whether the variability exhibited in the s-chart "explains" the ob-

Table 11.1 Control Chart Data Involving Time

Time interval	Average	Standard deviation	Time interval	Average	Standard deviation
1	69.76	1.87	16	72.61	3.79
2	71.32	2.82	17	71.69	3.86
3	73.93	5.06	18	74.69	2.98
4	69.95	3.58	19	69.40	1.58
5	72.93	3.11	20	71.24	4.75
6	70.65	3.99	21	71.26	3.57
7	74.78	1.88	22	71.54	4.56
8	73.32	1.72	23	74.36	3.31
9	72.16	2.12	24	75.16	3.54
10	71.86	1.16	25	73.56	3.25
11	72.37	1.74	26	75.06	3.06
12	71.05	4.70	27	72.11	4.21
13	71.76	4.47	28	74.99	2.09
14	68.49	2.57	29	71.67	2.50
15	70.08	6.28	30	71.94	5.37

Grand Average = $\bar{\bar{x}}$ = 72.1897
Average Standard Deviation = $\bar{s}$ = 3.3163

served variability of the $\bar{x}$: if all points fall inside the control band, the answer to this question is in the affirmative. The main reason for control chart work is to study the future in terms of the base period. For that purpose, it is what occurs *after* the base period that matters most. Extension of the central and control lines beyond the base period are the means to make that study. In our example, the study was continued for another ten points. It is clear from the graph that the points were no longer "in control" during this added phase of the experiment, and to detect such events is exactly the reason for running the control chart. The charts tell us that the process requires remedial action at this point. One of the very desirable features of the $(\bar{x}, s)$ control charts is that it allows us to study changes in level as well as in variability.

The control lines in this example were drawn at "± 2 sigma

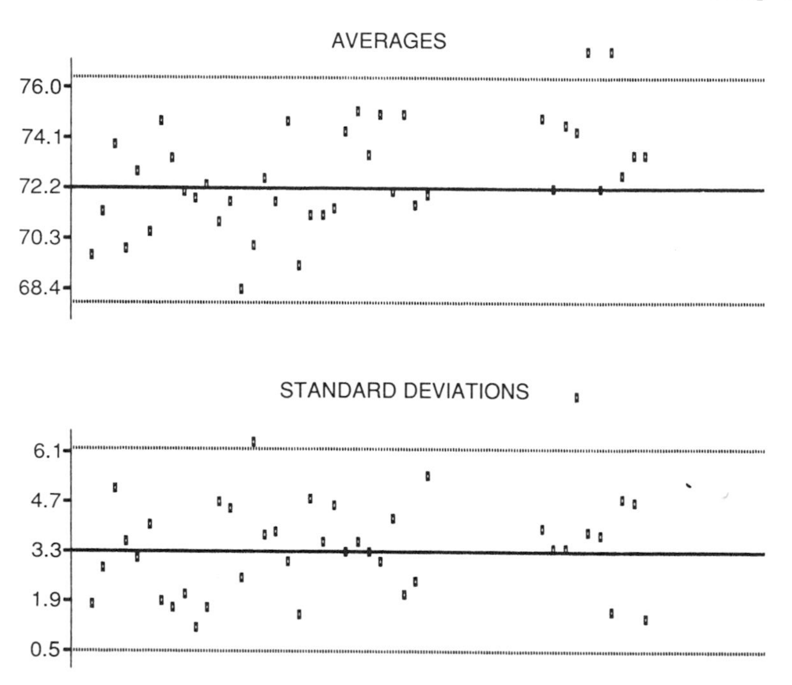

Figure 11.1 Control charts: averages and standard deviations.

deviations" from the mean. This corresponds to a control band containing approximately 95 percent of all values, for normally-distributed variables. However, the benefit of control charts derives, not from the 95 percent probability value, but rather from the empirical observation that the chart, with these limits, generally accomplishes the objective

Table 11.2 Ten Additional Control Chart Points

Time interval	Average	Standard deviation	Time interval	Average	Standard deviation
1	74.82	3.94	6	72.25	3.79
2	72.26	3.34	7	77.40	1.66
3	74.64	3.40	8	72.75	4.73
4	74.33	7.56	9	73.55	4.70
5	77.40	3.80	10	73.49	1.49

of maintaining the process at an approximately constant level. It is perfectly acceptable to draw the control lines at a different level; the optimum place is at the discretion of the user who, by experimentation, must determine the location that serves him best. Control charts were conceived by Shewhart as an empirical tool, rather than one based entirely on theory. It is indeed a very powerful tool.

11.3 A SECOND EXAMPLE

The second example is a study of randomization. Two thousand sheets of paper were to be arranged in groups of ten each. Each group of ten was to be sent to a different laboratory. The purpose was to compare the performance of the laboratories in testing the sheets.

Now, the initial stack of 2000 sheets was taken from production and there could very well exist a trend in the quality of the sheets. Thus, it is necessary to thoroughly randomize the sheets prior to allocation to the laboratories.

A computer study was made to simulate the randomization which was to be performed by means of a collator, as follows.

The sheets are stacked in the collator in five groups of 2000/5, or 400 each. The collator grabs one sheet from each group and then makes stacks of five each. The stacks are randomized, without disturbing the order of the sheets in each stack, and reassembled in this random order. They are reintroduced into the collator, now in seven groups of approximately 2000/7, or 285 each. The collator now produces stacks of seven sheets each. These are randomized, and the process repeated, using 11 groups of 2000/11, or 182 each. At the end, the entire stack is divided, in the order obtained, in consecutive groups of 10 sheets each.

It is possible, in the computer simulation, to keep track of the original number of each sheet (1 to 2000), in the order of production. The data in Tables 11.3 and 11.4 are averages of 10 such numbers. They are arranged in 15 groups of four. For each group of four, we calculated the average and the standard deviation. These are also listed in the tables. Tables 11.3 and 11.4 are only the first 15 of a larger number of sets.

For Table 11.3, the results are those obtained after only one pass

Table 11.3 Randomization Experiment: One Pass

	Group					Standard
	1	2	3	4	Average	deviation
1	968.50	1097.50	1117.00	984.50	1041.8750	76.1877
2	1064.00	983.50	921.50	1040.00	1002.2500	63.5354
3	949.50	1067.50	968.00	1015.00	1000.0000	52.7747
4	945.50	907.50	1106.50	950.50	977.5000	88.1174
5	915.00	1086.00	1006.00	1082.00	1022.2500	80.4171
6	915.00	985.00	871.50	1121.50	973.2500	109.3347
7	964.00	1027.00	1086.00	985.50	1015.6250	53.7112
8	1085.00	955.00	992.00	943.00	933.7500	64.3085
9	1050.50	968.00	854.50	1016.00	972.2500	85.4795
10	1053.50	1057.50	947.00	980.50	1009.6250	54.7333
11	876.50	867.50	932.00	962.50	909.6250	45.3438
12	831.50	889.00	983.50	972.00	919.0000	71.9386
13	1162.50	1088.50	1040.50	929.00	1055.1250	97.9195
14	955.00	1018.00	1034.50	1023.00	1007.6250	35.7570
15	1004.50	975.00	1050.00	986.00	1003.8750	33.0716

(from stacks of five sheets, each), followed by subdivision into groups of ten sheets.

Table 11.4 gives similar results, after three passes (stacks of 5, 7, 11 sheets), followed by subdivision into groups of 10 sheets.

Figures 11.2 and 11.3 are control charts for Tables 11.3 and 11.4, respectively.

The theory of the process is straightforward. Assume that there are N sheets in total. Then the average sheet number is $(N + 1)/2$. Let x be the original number of a sheet. If we assume complete randomization, the variance, for the set of N sheets, is

$$\mathrm{Var}(x) = E(x^2) - (Ex)^2$$

$$= \frac{\Sigma x^2}{N} - \left(\frac{\Sigma x}{N}\right)^2 = \frac{N(N + 1)(2N + 1)}{6N} - \left(\frac{N(N + 1)}{2N}\right)^2$$

$$= \frac{N^2 - 1}{12}$$

Table 11.4 Randomization Experiment: Three Passes

	Group					Standard
	1	2	3	4	Average	deviation
1	1240.10	1423.80	939.90	1007.20	1152.7500	221.8018
2	1071.70	1206.50	965.70	1023.80	1066.9250	102.6487
3	876.50	1110.00	1018.10	848.70	963.3250	122.7346
4	1048.10	580.00	1101.90	1114.70	970.1750	260.4191
5	1315.30	1042.00	917.70	1451.00	1181.5000	244.6653
6	993.50	811.80	1480.50	757.50	1010.8250	328.9779
7	708.80	1155.50	1125.50	598.40	897.0501	284.9653
8	1062.00	1021.10	974.50	968.10	1006.4250	43.9392
9	1123.10	944.70	714.70	844.60	906.7750	172.2337
10	974.50	977.30	1064.80	959.80	994.1001	47.7542
11	1140.00	1004.20	967.60	1097.20	1052.2500	79.9896
12	837.90	1323.40	980.40	899.40	1010.3250	216.7443
13	964.50	1162.50	721.80	1201.30	1012.5250	219.8150
14	1271.80	935.00	918.40	1101.40	1056.6500	165.5334
15	722.90	907.90	901.90	1042.20	893.7251	131.0137

For stacks consisting of 10 sheets, the average per stack is still $(N + 1)/2$, and the standard deviation is

$$\sigma_{\bar{x}} = \sqrt{\frac{N^2 - 1}{120}}$$

From each group of four averages of ten, calculate a standard deviation, with three degrees of freedom. Denote the standard deviation by $s_{\bar{x}}$. Then:

(a) $s_{\bar{x}}$ is an estimate of $\sigma_{\bar{x}}$

(b) the standard deviation of $s_{\bar{x}}$ is approximately (see Eq. 10.7):

$$s(s_{\bar{x}}) = \frac{\sigma_{\bar{x}}}{\sqrt{2(4 - 1)}} = \sqrt{\frac{N^2 - 1}{720}}$$

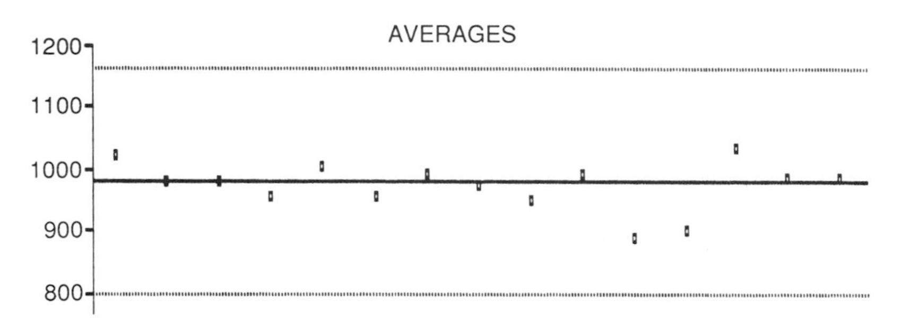

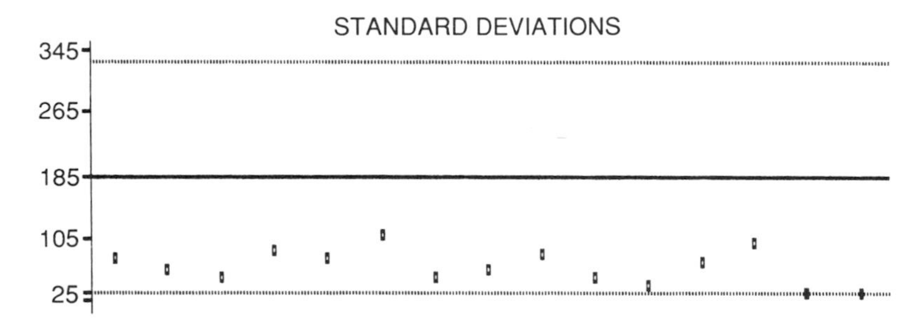

Figure 11.2 Randomization experiment. One pass.

We now have all the quantities required for a control chart for averages and for standard deviations. For the chart for averages

$$\begin{cases} \text{the central line is at } (N+1)/2 \\ \text{the 2-sigma control lines are at} \\ (N+1)/2 \pm 2\sqrt{(N^2-1)/(120 \times 4)}^{*} \end{cases}$$

For the chart for standard deviations

$$\begin{cases} \text{central line is at } \sqrt{(N^2-1)/120} \\ \\ \text{2-sigma control lines are at } \sqrt{(N^2-1)/120} \pm \sqrt{(N^2-1)/720} \end{cases}$$

For $N = 2000$, we thus have:

*The factor 4 is necessary, because the average plotted is the average of a group of four individual averages.

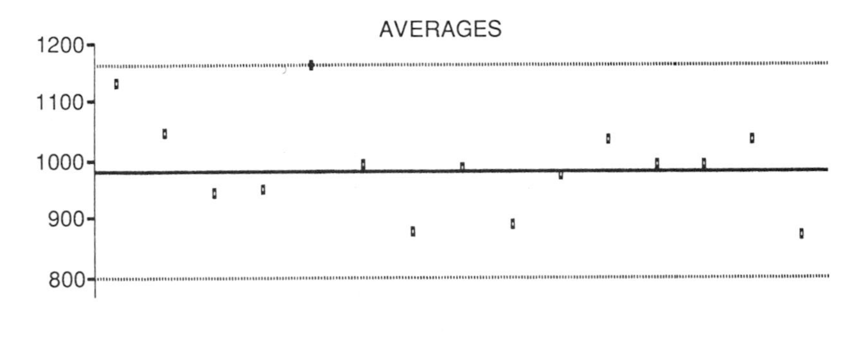

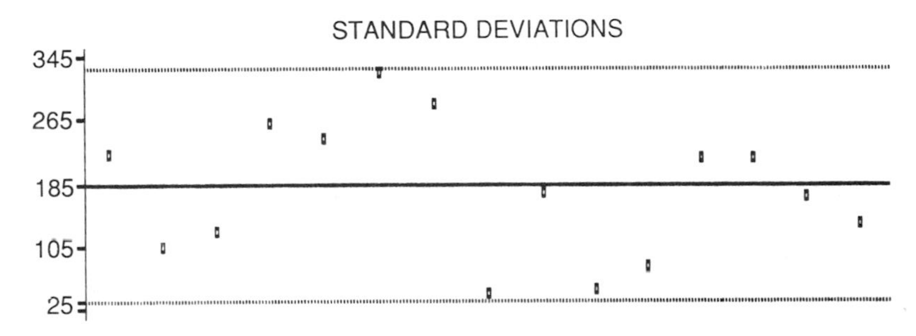

Figure 11.3 Randomization experiment. Three passes.

Chart for averages:

$$\begin{cases} \text{the central line at 1000.5} \\ \text{the 2-sigma control lines, at 817.92 and 1183.07} \end{cases}$$

Chart for standard deviations:

$$\begin{cases} \text{central line at 182.57} \\ \text{2-sigma control lines at 33.50 and 331.64} \end{cases}$$

An examination of Tables 11.3 and 11.4 and of Figures 11.2 and 11.3 shows very clearly that whereas one pass does *not* lead to satisfactory randomization, three passes, with bins of 5, 7, and 11, result in data that are entirely consistent with the theoretical predictions based on complete randomization. Indeed, the one-pass control chart for averages shows values that are too close to the grand average, whereas the chart of standard deviations exhibits points that are *all* well below the expected average. Both these facts are consistent with the presence of strong positive correlation between the observed averages, indicating failure to obtain randomness. On the other hand, with three passes, the control charts behave very much in accordance with predictions from an assumption of completely randomized sheets.

This example is a somewhat unorthodox application of control chart practice. It shows, as Wernimont (1946) pointed out long ago, that control charts can be effectively used as a scientific tool to solve practical problems of many different types.

11.4 CONTROL CHARTS FOR DISCRETE DATA

Many physical or chemical tests do not result in continuous measurements, but rather in go–no-go verdicts. This is true whenever a test results is a verdict of *conforms* or *does not conform,* after comparison of the result with a specification.

If items are taken at random from a process which produces a fraction P of nonconforming (*defective*) articles, and a fraction $Q(=1-P)$ of conforming articles, and one wishes to judge whether the process is stationary, i.e., not-changing, a different type of control chart is required.

Suppose that N experiments are performed, and that every experiment consists of examining n items for conformance. The results are represented by $a_1, a_2, \ldots, a_N$, where a_i is the number of nonconforming articles in the ith experiment. Then, in the long run (N very large) we will have:

$$\text{Average of } a_i/n = P$$

It can also be shown that the variance of a_i/n will be

$$\text{Variance of } a_i/n = P\frac{Q}{n}$$

In terms of a_i (as opposed to a_i/n) we have:

$$\text{Average of } a_i = nP$$

$$\text{Variance of } a_i = nP \cdot Q$$

The distribution curve governing data of this type, called the "binomial distribution," is given by the equation

$$\text{Prob}(n,m) = C_n^m P^m Q^{n-m}$$

Prob(n,m) is the probability of obtaining exactly m nonconforming items in a sample of size n; C_n^m is the number of combination of n articles m at a time, and P and Q are as defined above.

The formula for C_n^m is

$$C_n^m = \frac{n!}{m!\,(n - m)!}$$

where the symbol ! denotes a factorial. Thus, $m! = 1 \cdot 2 \cdot 3 \cdots$ m.

Note that we are dealing with *a discrete* (noncontinuous) distribution: m starts at zero and increases by steps of unity: m = 0, 1, 2, etc. It cannot be fractional. (For $m = 0$, C_n^0 is defined as $C_n^0 = 1$). As an example, let $n = 10$ and $P = 0.08$. Then the probability of getting $m = 0, 1, 2, \ldots$ nonconforming items in a sample of 10 is $C_{10}^m P^m Q^{10-m}$. The values for this probability are given in Table 11.5, assuming $P = .08$ for m = 0 to $m = 8$. In this case, Prob(n,m) is steadily decreasing. However, for larger values of n, (starting with $n = 12$ for $P = 0.08$), the value of Prob(n,m) will first increase, reach a maximum, and then decrease. The distribution, for small n, is very skew (asymmetrical). For large n it becomes more symmetrical. Note that Prob(n,m) is also the probability of obtaining a *fraction* m/n of nonconforming items in a sample of n items. For large n, the steps $0/n$, $1/n$, $2/n$, etc., become smaller and smaller, and the distribution expressed in terms of this fraction, becomes more and more continuous. It can be shown that it approaches normality for large n.

We now discuss a numerical example obtained by computer-simulation. The first four parts of Table 11.6 exhibit the data for 40

Table 11.5 Probability of m Non-Conforming Items in a Sample of Size 10, for $P = 0.08$

m	m/n	Probability
0	0	.434388
1	.1	.377729
2	.2	.147807
3	.3	.034274
4	.4	.005216
5	.5	.000544
6	.6	.000039
7	.7	.000002
8	.8	.000000

sets (N = 40) of size 10 each (n = 10). The value of P is 0.08. Thus, the example could express a situation in which a manufacturing process (for example, of transistors) produces eight percent of *defective* (non-conforming) transistors in the long run. We imagine that we sample 40 consecutive shifts, by taking a sample of 10 transistors from each shift, and count the number of defective items in each sample of 10.

Table 11.6 Fraction Defective in 40 Sets of Size 10 when $P = 0.08$ Followed by 10 Sets when $P = 0.28$

Set	Fraction defective	Set	Fraction defective	Set	Fraction defective	Set	Fraction defective	Set	Fraction defective
1	.1	11	.1	21	.4	31	0	41	.2
2	.1	12	0	22	.1	32	.1	42	.4
3	0	13	0	23	0	33	.2	43	.5
4	.1	14	.2	24	.1	34	.3	44	.3
5	0	15	0	25	0	35	.1	45	.3
6	.1	16	.1	26	.1	36	.1	46	.2
7	0	17	0	27	.1	37	0	47	.2
8	.4	18	0	28	0	38	0	48	.4
9	.2	19	.1	29	.1	39	.1	49	.2
10	0	20	.1	30	0	40	.1	50	.2

We also suppose that after 40 shifts, the process is suddenly adversely affected, producing now 28, rather than eight percent of defective items (last part of Table 11.6). We wish to be able to detect this adverse effect, and propose to do it by means of a control chart. We assume that P is not known to the inspector.

From an examination of the 40 sets of results, the inspector is able to *estimate* P, by simply averaging the fractions defective. Thus, the estimate p of P is

$$p = \frac{\Sigma_i a_i/10}{40}$$

where a_i is the number of defectives, observed in the ith sample. Then the standard deviation of the (a_i/n) values can be calculated from the equation:

$$s = \sqrt{\frac{p(1 - p)}{n}} = \sqrt{\frac{p(1 - p)}{10}}$$

For Table 11.6, we find for the first 40 sets, $p = .0875$ and consequently $s = .0894$.

Some textbooks recommend that a control chart be drawn with a central line at $\bar{p}$ and three-sigma control limits at $\bar{p} \pm 3s$. When $\bar{p} - 3s$ is negative (as it is in our case) it is replaced by zero. Such a chart is shown on Fig. 11.4. The calculations are made on the first 40 points but the chart is extended for 10 additional points for which P = 0.28. Clearly, the graph is very unsatisfactory. Two points out of 40 are above the upper control line. There are 16 points at zero, the lower control line. Yet ± 3 sigma limits should include about 99.7% of all points. The reasons for this behavior are twofold: the binomial is a *discrete* distribution and thereby presents problems in drawing control lines at given probability values. This is further discussed below. Secondly, the binomial is, for small n, quite different from the normal distribution, for which three-sigma limits have the probability meaning usually attributed to them.

Table 11.5 shows what the correct population value of P, in this case $P = 0.08$, yields.

We see that the probability for a fraction defective equal to zero is 0.4345. In other words: about 43 percent of all points should show

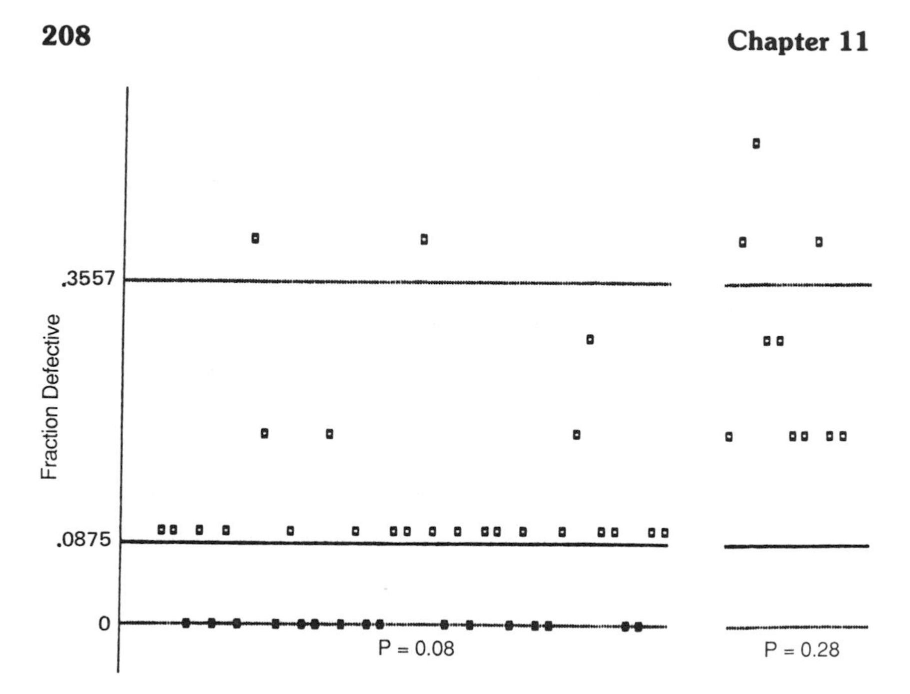

Figure 11.4 Control chart for binomial original scale.

a fraction defective equal to zero. It is therefore impossible to draw a lower control line that would *exclude* 0.14% of the values, (which the three-sigma control line for normally distributed variables does). Similarly, if we draw the upper control line above m = 4, we exclude only 0.05% of all points. If we draw it between $m = 3$ and $m = 4$, we exclude 0.58% of all points. Here again we are to satisfy an impossible demand.

The control chart in question is misleading and should never be made. What can be done? It has been shown that if we transform the individual p values to

$$z = \arcsin \sqrt{p}$$

(this is called the *arcsine transformation*) then the z values are much more closely approximated by the normal distribution than the original p-values. The approximation is further enhanced if, for $n<50$, we replace every zero value of p by $n/4$ and every $p = 1$ value by $(n -$

0.25)*n*. These substitutions must be made *prior* to the arcsine transformation. The average, in the arcsine scale, should be close to arcsin $\sqrt{P}$. The standard deviation, in the arcsine scale, is given by

$$\sigma = \frac{0.5}{\sqrt{n}}$$

For our example, we have

$$\arcsin \sqrt{P} = \arcsin \sqrt{.08} = 0.2868$$

$$\sigma = \frac{0.5}{\sqrt{10}} = .1581$$

We find, for the actual average of our arcsine values: $\bar{x} = .2918$ and for their standard deviation: $s = .1414$.

If the calculated lower control line, $\bar{x} - 3s$, is negative, it is replaced by zero.

Figure 11.5 exhibits the 3 sigma control chart for the arcsine

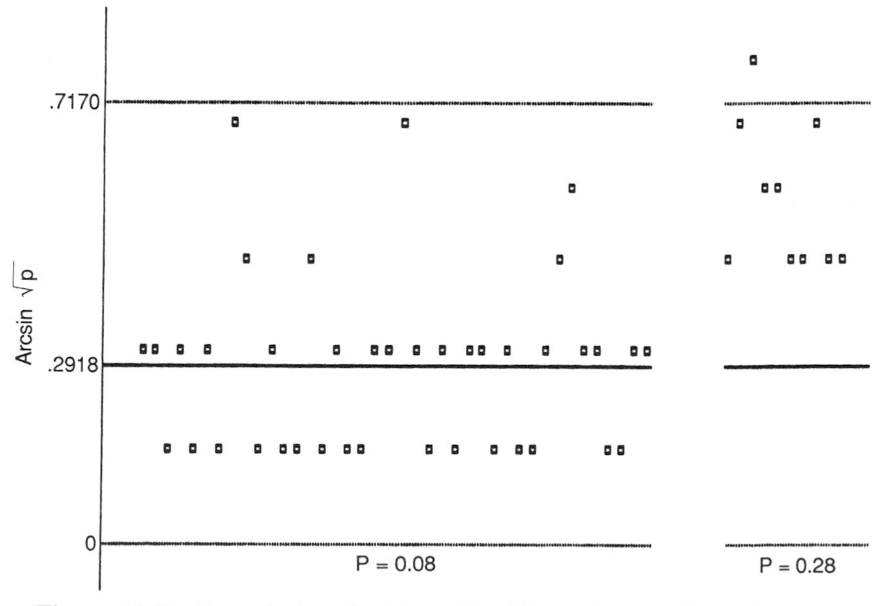

Figure 11.5 Control chart for binomial with arcsine transformation.

values. The central line is at 0.2918, the lower control line at zero, and the upper control line at 0.7160 (= .2918 + 3 × .1414). The graph is certainly better than that obtained for the untransformed data. It is also seen that the sudden change in P, after 40 points, is apparent on the graph.

It is important to observe that the arcsine transformation does not eliminate the discrete nature of the data: the values now have different "labels" (arcsin $\sqrt{x}$ instead of x). But the distribution of the values in the arcsine scale is closer to the normal distribution than that of the original values.

In conclusion, it seems advisable to plot binomial results in an (arcsin $\sqrt{x}$) scale. A marked change in the probability of defectives, *P*, is apparent in the graph. This can be used for controlling the stability of a binomial process.

11.5 CONCLUSION

The control chart is a simple, but powerful tool for the analysis of data. It is also an effective way of controlling the quality of measurements over a period of time. We have shown that it can also be used with great effectiveness for the solution of special problems arising in science and in quality control.

REFERENCES

Shewhart, Walter, A. (1931). Economic Control of Quality of Manufactured Product, Van Nostrand, NY.

Wernimont, Grant (1946). Use of Control Charts in the Analytical Laboratory. Industrial and Engineering Chemistry, Analytical Edition, *18*, 586–592.

12

Comparison of Alternative Methods

12.1 THE PROBLEM

It often happens that the same physical or chemical property can be measured in different ways. For example, one can determine sodium in serum by flame atomic emission spectroscopy or by isotope dilution-mass spectrometry. The question then arises as to which method is *better*.

Our first task is to define better. As we mentioned earlier, accuracy is not a major concern, since we can generally calibrate a method in terms of another one. Therefore, we are mostly interested in the precision of the method. However, precision is generally measured by a standard deviation, in units of the measurement. Thus, when we measure sodium in serum by flame atomic emission spectrometry, the result is the relative intensity of a spectrometric peak. The problem then is to compare methods that are expressed in different units.

12.2 THE CRITERION

Let x and y represent the two methods. Consider a particular value of x, say x_0. The corresponding value of y is y_0. In a small interval near x_0, say x_0 to $x_0 + \Delta x$, the curve of y versus x can be considered straight. The corresponding y-interval is y_0 to $y_0 + \Delta y$. The ratio $\Delta y/\Delta x$ represents the slope of the line near x_0 (Figure 12.1). If we wished to express y in units of x, we would multiply y by $\Delta x/\Delta y$. Thus, the standard deviation of y, near the point (x_0,y_0) would become $\sigma_y \cdot \Delta x/\Delta y$. This is the value that must be compared to σ_x. If $\sigma_y \cdot \Delta x/\Delta y$ is considerably smaller than σ_x, then y is *better* (more *sensitive*) than x. We therefore consider the quantity

$$S_{y/x} = \sigma_x \Big/ \left(\sigma_y \frac{\Delta x}{\Delta y}\right)$$

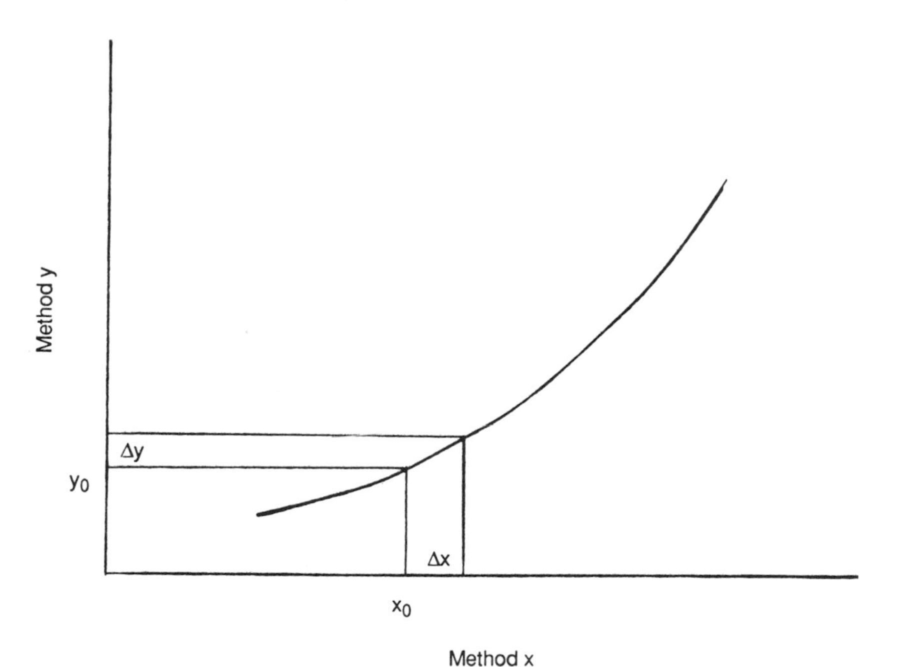

Figure 12.1 The sensitivity criterion.

which can be written in the form

$$S_{y/x} = \frac{\Delta y/\Delta x}{\sigma_y/\sigma_x} \tag{12.1}$$

which we call the *sensitivity of y with respect to x*. We see that the sensitivity is the quotient of the slope by the ratio of standard deviations. A value of S considerably *larger* than unity indicates that y is more sensitive (better) than x. A value of S considerably *smaller* than unity indicates that x is more sensitive.

It is important to realize that S is a *local* quantity: the slope $\Delta y/\Delta x$ may well vary from one point to another, and we have already seen that both σ_x and σ_y may be functions of the level. Therefore, the sensitivity S must be measured at a sufficient number of points to express it as a function of the point on the curve.

12.3 EXPERIMENTAL DETERMINATION OF SENSITIVITY

An experiment to determine the sensitivity of a method y with respect to a method x is readily conceived.

Consider a number of materials for which the values of x and y are

$$\begin{array}{ccccc} x_1 & x_2 & x_3 & \cdots & x_p \\ y_1 & y_2 & y_3 & \cdots & y_p \end{array}$$

By plotting y versus x we obtain an approximation to the curve y versus x. If, for each of the p materials we make a number of replicate measurements by the x method and a number of replicate measurements by the y method, we will be able to approximate, at each point, both σ_x and σ_y. The slope of the line at each point is obtained once we are able to fit an equation to the x,y relation.

12.4 A NUMERICAL EXAMPLE

An experiment was performed to evaluate two methods for measuring the quality of dried eucalypti veneers (Kauman, Goldstein, and Lau-

Table 12.1 Quality of Dried Eucalypti Veneers: Results of Numerical Quality Evaulation

	Quality ratings					
	Observer A		Observer B		Observer C	
Sheet no.	Test no. 1	Test no. 2	Test no. 1	Test no. 2	Test no. 1	Test no. 2
1	22	15	21	24	19	11
2	31	23	24	21	29	33
3	34	37	32	35	40	37
4	44	44	41	44	42	40
5	10	6	13	19	8	7
6	31	32	34	39	34	25
7	42	39	39	41	25	30
8	28	19	23	27	13	24
9	29	29	27	27	28	25
10	39	37	29	35	34	36
11	21	20	17	25	16	23
12	42	44	44	40	41	38
13	40	39	34	37	34	31
14	11	9	9	12	7	8
15	38	37	30	36	44	40
16	9	13	14	25	11	12
17	20	16	19	19	6	15
18	28	31	30	31	24	37
19	28	24	25	30	25	22
20	15	8	11	19	9	10
Mean	28.1	26.1	25.8	29.3	24.4	25.2

tican, 1956). One was numerical, the other subjective. Tables 12.1 and 12.2 show the results. A low rating is "good," a high rating "bad." We wish to compare the two methods. Using the method of Chapter 10, we first determine the reproducibility of both sets of data. The results are shown in Table 12.3. There is no evidence of a relation between the standard deviations and the value of the rating for either method. The pooled values of the standard deviations are given at the

Table 12.2 Quality of Dried Eucalypti Veneers: Results of Subjective Quality Evaulation

	Quality ratings					
	Observer A		Observer B		Observer C	
Sheet no.	Test no. 1	Test no. 2	Test no. 1	Test no. 2	Test no. 1	Test no. 2
1	3	2	3	4	3	4
2	7	7	5	4.5	7	5
3	6	5	5	5	5	6
4	7	7	8	8	7	6
5	1	1	2	2	3	2
6	5	5	5	6	5	5
7	5	6	6	6	5	5
8	3	3	5	4	4	4
9	4.5	4	5	5	5	5
10	6	6	7	6	7	7
11	5	4	4	4	4	5
12	8	7	7	8	8	6
13	5	6	7	7	5	5
14	1	1	2	2	2	3
15	7	8	7	7	7	8
16	1	3	3	3	4	4
17	4	3	3	3	3	4
18	6	4	6	5	5	5
19	5	4	5	5	7	5
20	3	2	3	4	2	2
Mean	4.62	4.40	4.90	4.92	4.90	4.80

bottom of the table. The ratio of the standard deviations, subjective to numerical, is

$$0.8088/4.5152 = 0.1791$$

The averages of Table 12.3 are plotted against each other in Figure 12.2. There is no evidence of curvature.

Table 12.3 Summary of Analysis of Tables 12.1 and 12.2

	Numerical		Subjective	
Veneer	Average	Standard deviation of reproducibility	Average	Standard deviation of reproducibility
14	9.333	1.936	1.833	0.816
5	10.500	5.212	1.833	0.816
20	12.000	4.359	2.667	0.866
16	14.000	5.852	3.000	1.155
17	15.833	5.447	3.333	0.577
1	18.667	4.924	3.167	0.764
11	20.333	4.359	4.333	0.577
8	22.333	6.028	3.833	0.816
19	25.667	2.887	5.167	1.000
2	26.833	5.050	5.917	1.275
9	27.500	1.581	4.750	0.456
18	30.167	5.462	5.167	0.913
6	32.500	4.682	5.167	0.408
10	35.000	3.559	6.500	0.577
3	35.833	2.930	5.333	0.577
13	35.833	3.731	6.000	0.957
7	36.000	7.577	5.500	0.577
15	37.500	4.967	7.333	0.577
12	41.500	2.380	7.333	1.000
4	42.500	1.826	7.167	0.816
Pooled Standard Deviation		4.5152		0.8088

Using the method of straight line fitting explained in Section 5.5, with $\lambda = (0.1791)^2$, we obtain:

Subjective = 0.4781 + 0.1619 · numerical

Thus, the sensitivity ratio, subjective to numerical, is

$$S_{S/N} = \frac{.1619}{.1791} = .9093$$

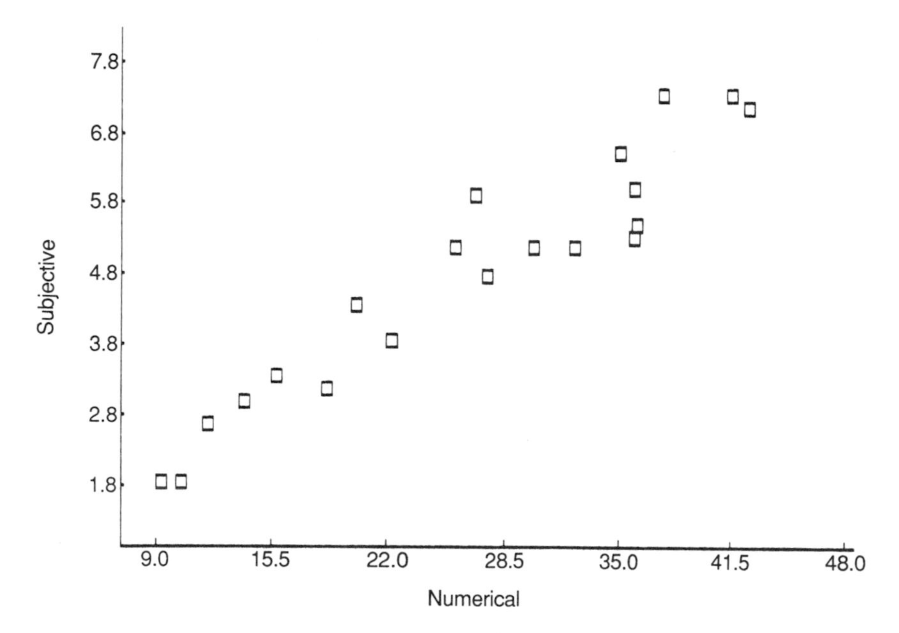

Figure 12.2 Veneer date. Subjective versus numerical.

We see that the ratio is quite close to unity, indicating about equal sensitivity for both methods.

12.5 ECONOMIC AND TECHNICAL MERIT. SAMPLE SIZE

Suppose that the sensitivity of method A with respect to method B is

$$S_{A/B} = \frac{\Delta A/\Delta B}{\sigma_A/\sigma_B}$$

This can be written as:

$$S_{A/B} = \sigma_B / \left(\sigma_A \cdot \frac{\Delta B}{\Delta A} \right)$$

If $S_{A/B} > 1$, then, by making n measurements of B, we can choose n in such a way that

$$\frac{\sigma_B}{\sqrt{n}} = \sigma_A \cdot \frac{\Delta B}{\Delta A} \tag{12.2}$$

Then, n measurements by B will be equivalent, in terms of sensitivity, to a single measurement by A. Equation (12.2) can be written:

$$n = \left(\frac{\Delta A/\Delta B}{\sigma_A/\sigma_B}\right)^2 = S^2_{A/B} \tag{12.3}$$

Equation (12.3) shows that the sample size n required to make the sensitivity of the less sensitive method equal to that of the other method is the square of the sensitivity ratio. Now, assume that a single measurement by A costs a dollars, and a single measurement by B costs b dollars. Then, in order to achieve the same sensitivity, one will have to spend bn dollars by method B for a dollars by method A. Thus, for equivalent precision, the ratio of the costs will be

$$\text{ratio} = bn/a \tag{12.4}$$

For example, if $S_{A/B} = 2.5$, $a = \$3.00$ and $b = \$0.48$, then according to Eq. (12.3) we have

$$n = (2.5)^2 = 6.25$$

Hence, the ratio of the costs (B to A) is:

$$0.48 \times 6.25/3 = 1.00$$

In other words, it will take a little over six measurements by B to reach the same precision as would be obtained by a single A measurement. However, because B is cheaper, the cost would be the same for both methods. We see that the *economic merit* of a test method, measured by its cost, can be combined with its *technical merit,* measured by its sensitivity, to make rational decisions in choosing a test method.

12.6 INVARIANCE OF THE SENSITIVITY RATIO UNDER SCALE TRANSFORMATIONS

The results of any measurement instrument can be expressed in different scales. Today, with the automation of instruments, all that is required is to install a reading dial in any desired scale. Thus, a spectrophotometer could have a scale expressed in *percent transmittance*; but it could also have a scale expressed in *optical density*. Thermometers can be read in degrees centigrade, Fahrenheit, or Reaumur. Acidity can be expressed in concentration of hydrogen ions, on in pH units.

In some cases, the alternate scales are proportional to each other (e.g., centimeters and inches), in others there is a linear relation, but no proportionality (e.g., degrees centigrade and Fahrenheit). In many cases the relation is not even linear (e.g., hydrogen ion concentration and pH).

The intrinsic precision of a measuring device should not be affected by the scale used in any particular application, except for the readability of the scale by the observer. A true measure of precision should therefore be independent of the particular scale used, even when nonlinearly related scales are compared. The sensitivity, as defined above, is such an invariant measure. The proof is very simple.

Suppose that two different methods, A and B, are compared. Let $S_{A/B}$ represent the sensitivity of A with respect to B

$$S_{A/B} = \frac{dA/dB}{\sigma_A/\sigma_B}$$

Since the slope could be negative, a more appropriate definition is

$$S_{A/B} = \frac{|dA/dB|}{\sigma_A/\sigma_B}$$

Now suppose that the scale of A is changed to A^* so that $A^* = f(A)$ where $f(A)$ is a continuous function. For example pH $= -\log_{10}$ (hydrogen ion concentration). For this example, $f(A) = -\log_{10}(A)$.

In the general case, the slope of A to B becomes

$$dA^*/dB = \frac{dA^*}{dA} dA/dB \tag{12.5}$$

But dA^*/dA is the first derivative of A^* with respect to A, or $f'(A)$.

Furthermore

$$\sigma_{A^*} = \left|\frac{dA^*}{dA}\right|\sigma_A = |f'(a)|\sigma_A \tag{12.6}$$

Thus,

$$S_{A^*/B} = \frac{|dA^*/dB|}{\sigma_{A^*}/\sigma_B}$$

which, according to Eqs. (12.5) and (12.6), becomes

$$S_{A^*/B} = \frac{|f'(a)dA/dB|}{|f'_A|\sigma_A/\sigma_B} = \frac{|dA/dB|}{\sigma_A/\sigma_B}$$

Similarly, it is shown that a change of scale in B leaves the sensitivity ratio unchanged.

This invariance property of the sensitivity ratio is of the greatest importance. Contrary to widespread belief the coefficient of variation (standard deviation/value of measurement) does *not* enjoy invariance.

Suppose, for example, that a thermometric measurement in degrees centigrade, say A, is compared to a different measurement of temperature, say B. Then the ratio of the coefficients of variation is

$$\frac{\sigma_A/A}{\sigma_B/B}$$

Now change the scale of A to degrees Fahrenheit, say A^*. Then since

$$A^* = 32 + 9/5A$$

We have

$$\sigma_{A^*}/A^* = 9/5\alpha_A/(32 + 9/5A)$$

and the ratio becomes:

$$\frac{\frac{9}{5}\sigma_A \Big/ \left(32 + \frac{9}{5}A\right)}{\sigma_B/B}$$

This is by no means the same as $\sigma_A/A/\sigma_B/B$. Note that what spoils the invariance is the constant 32. If the two scales were *proportional* to each other, and only then, would the coefficient of variation be an invariant measure.

One should resist the temptation to compare methods of test *on a relative basis,* i.e., by comparing their coefficients of variation, for the reasons just explained.

12.7 SENSITIVITY AS A BASIC MEASURE OF UNCERTAINTY

The science of statistics deals with populations. These have fixed means and fixed standard deviations. For distributions that are close to Gaussian, the standard deviation is an accepted measure of uncertainty. Consider the result of a measurement x. It is generally considered that

$$x = E(X) + \varepsilon \tag{12.6}$$

where $E(X)$ is the population mean of x. The term ε measures the deviation of an observed measurement x from its population value $E(X)$. The standard deviation of ε, σ_ε, is a measure of the uncertainty of x.

When the measurement process x is compared to another process say y, then the comparison of σ_x to σ_y is of paramount interest. Such a comparison makes sense when $E(x)$ and $E(y)$ are essentially equal to each other. For example, x can be the result of a gravimetric determination of the manganese content of an ore, and y can be the result of spectrophotometric measurement for manganese on the same ore, both expressed as *percent manganese*. But modern statistical test books deal elaborately with transformation of scale, for a variety of reasons. If either x or y are expressed in a transformed scale, the comparison of σ_x and σ_y is no longer meaningful. Suppose, for example, that a square root transformation is made on x: $z = \sqrt{x}$. If the error ε, of x, is relatively small, then the standard deviation of z will be $\sigma_z = \sigma_\varepsilon/2$. Thus by transforming x to its square root we make the process x appear much more favorable in terms of precision. This confusing state of affairs is totally obviated if we consider, not σ_x, by the sensitivity of x as a measure of precision.

In order to consider the sensitivity of a measurement process x, it is necessary to select one of the alternative methods for determining the property in question as a *standard*. Let y be the selected standard method. The sensitivity of x with respect to y is a true measure of the merit of the process x: the larger this sensitivity the more we credit x as a precise method.

In dealing with sensitivity measures, more meaningful results will be obtained when the errors of measurement are relatively small (say, not exceeding 10% of the measured value). Our example (Section 12.4) does not satisfy this requirement. Therefore, the final conclusion is not very firm in this case. Most measurements in a chemical laboratory will lead to conclusions of greater reliability.

12.8 CONCLUSION

It is possible to compare methods of testing even when their results are expressed in different scales. A criterion called sensitivity has been presented for accomplishing this goal. When comparing methods of testing, considerations based on economic factors must be taken into account in conjunction with the results obtained through the application of the criterion of sensitivity.

REFERENCE

Kauman, W. G., J. W. Goldstein and D. Lautican (1956). Quality Evaluation by Numerical and Subjective Methods with Application to Dried Veneer, Biometrics, *12*, 127–153.

13

Data Analysis: Past, Present, and Future

13.1 INTRODUCTION

It has been often stated that statistics is applied probability theory. Certainly this is still tacitly assumed, though not always clearly stated. Our textbooks stress the importance of tests of significance and confidence intervals, which are in a major sense statements of probability. Our journals abound in probabilistic developments, and worthwhile advances in statistics are generally regarded as requiring advances in sophistication of theoretical developments in applied probability theory.

Let us be specific by means of an example: Cramer (1974, p. 549) presents the problem of fitting a straight line as a problem of dealing with the normally distributed random variable y where a sample of n pairs is given such that $y_i = \alpha + \beta(x_i - \bar{x})$ for $i = 1$ to n. Note that y is a *random variable* depending on two parameters α and β.

The x-values are a vector of fixed numbers x_1 to x_n. The emphasis is on the probability aspect of y, and only incidentally on α and β.

12.2 MODELS

Today we would look upon this problem as one involving, first, a *model* such as

$$\eta = \alpha + \beta x \qquad \text{and} \qquad y = \eta + \varepsilon$$

where η is a linear mathematical function of x with parameters α and β, and, secondly, a random variable ε. The emphasis now is on two aspects: the model, and the nature of the random variable ε.

Dealing with a real-life problem, we would first try to verify the validity of the model (is the relation of η to x really linear?) and then deal with the random variable ε (is it normally distributed with variance independent of x; are the ε mutually independent?). We realize that both the model and the distribution of ε are probably not what we assume. We live in a world of unknowns, and can only make reasonable assumptions about the truth. These assumptions are very likely to be untrue, but they may be useful to us. Looked at from this viewpoint, probability statements based on *exact, true* assumptions are relatively unimportant to us. Moreover, we don't generally have the knowledge necessary for such statements. It is not unreasonable to assume that in the future, data analysis will be concerned more and more with useful approximations to the truth, with emphasis on the usefulness of inferences rather than on their probability values.

12.3 LOGICAL REQUIREMENTS

The above is not necessarily a pessimistic statement about data analysis. Indeed, data analysis, in order to be acceptable, must satisfy a number of logical requirements, lest it degenerate into mere arithmetic. The logical foundation of data analysis is, unfortunately, not adequately discussed in most textbooks. Furthermore, many recommended and widely used methods of data analyses are in flagrant violation of the logical basis underlying the field. Recommended practices have been published for dealing with data such as the results of interlaboratory studies of test methods that advance procedures that clearly violate the logic of inductive inference. What are the logical requirements for inductive reasoning?

1. First of all, the data must be given within the context of a given experimental procedure or method of observation. The statistician or data analyst should be given pertinent information about the procedure. He should never be given just a set of numbers with the request to *analyze them*. Background information is an essential part of what the analyst should be given to allow him to perform an intelligent analysis. For example, if data are presented in a two-way table with rows and columns, the precise meaning of these rows and columns in terms of the experiment that was performed, should be stated.

This does not mean that unnecessary and irrelevant information should be given. Here the judgment of the person or group of experts submitting the data comes in place. They should be thoroughly conversant with the subject matter so as to be able to judge these matters with competence.

2. Many experiments are not isolated events: often they are just one of a series of experiments carried out over a period of time. This raises the question of *previous information* influencing the inferences drawn from a single experiment. One method for taking previous information into account is to use Bayes' theorem (Press, 1989). However, with Bayes' theorem we are back in probability theory: the information available from previous experiments is included as a "prior probability." Many problems simply cannot be handled that way: they require specifications of models much more than probability statements. Proper data analysis must establish ways in which *series of experiments* can be analyzed. This matter has so far received little attention. It is an important item for the future of data analysis.

3. R. A. Fisher (1955) has drawn attention to a basic difference between *deductive* and *inductive* reasoning. Whereas in deductive reasoning (e.g., Euclidean geometry), conclusions can be drawn from certain premises, *regardless* of the validity of other premises, this is not the case with inductive reasoning. For example, it follows from the postulates of Euclidean geometry that the sum of the angles of a plane triangle is 180 degrees. This fact cannot be changed by additional

information, for example, about the lengths of the sides of the triangle. In inductive reasoning, on the other hand, a new, additional fact can radically change the conclusions. For example, if data are obtained in five laboratories, the subsequent addition of similar measurements in a sixth laboratory will change the estimates of repeatability and reproducibility. In other words, inferences should be based on the *entire* set of data, not on a portion of it. Here are Fisher's exact words: "In one respect inductive reasoning is more stringent than is deductive reasoning, since in the latter any item of the data may be ignored, and valid inferences may be drawn from the rest; i.e., from any selected subset of the set of axioms, whereas in inductive reasoning the whole of the data must be taken into account."

Most recommended practices dealing with interlaboratory test data recommend the use of Cochran's test, Dixon's test, or Grubbs' test for the rejection of outliers. These tests are based on the examination of the data for a single material, *and rejection rules are given on this basis*.

There is nothing wrong with looking at the data one material at a time, but after this is done it is *imperative* to study the results from the *entire* experiment, including *all* materials, *before* drawing final inferences. In our presentation of methods of analysis for interlaboratory data we have adhered to this precept: our h and k graphs are based on *all* the data, even though they were obtained by examining one material at a time.

4. In line with the previous point, it is also *impermissible* to ignore pertinent information. For example if a two-way table is presented, and the meaning of rows and columns is explicitly stated, then the data cannot be analyzed ignoring rows and columns and stringing the data out along a single, continuous line. This can only be done *after* it has been satisfactorily proved that the classification according to rows and columns does not affect the inferences drawn from the data, and is therefore irrelevant. Thus, how to look at the data is not entirely a matter of choice.

5. The human mind is generally incapable of understanding a large table of data: too many intercomparisons have to be made. On the other hand, humans are very good at understanding and interpreting graphs. Drawing the proper graphs is as yet not a science and it is

questionable whether it will ever be. But a good data analyst will give considerable attention to the question of what graphics he should make to answer the pertinent questions about a given data set. Again our h and k graphs (see Chapter 10) are examples of graphs that give almost instantaneous information on many questions pertinent to interlaboratory test data such as lab biases and outliers. The information given by these graphs is, moreover, quantitative without boggling the mind with numbers difficult to interpret.

There is no substitute for a graph for showing relationships between variables. A simple (x,y) plot shows at once what can be obtained only laboriously by analytical methods. For example, tests of significance on the intercept and the slope of a straight line, or on its correlation coefficient should *follow,* not precede the presentation of a graph. If the range of values is very large, different methods may have to be used to present the data graphically.

It is often necessary to make *several* graphs, highlighting different aspects of a set of data. Thus h and k refer to different aspects of interlaboratory data, h to between-laboratory differences, and k to within-laboratory variability. It is easily seen that h and k together do *not* exhaust the information provided by interlaboratory data. Neither one informs us as to the relationship between *level* and *variability*. Thus a third graph is necessary: a plot of repeatability and reproducibility standard deviation versus level.

6. In the last couple of decades, a field called "exploratory data analysis" has been advocated, taught, and propagated. Data analysis, according to this view, consists of two phases: exploratory and confirmatory. The distinction is not clear. It appears to this author that every analysis is exploratory, and that once a thorough exploration of the data has been made, there is little else left to do. The availability of high-speed computers has made exploration easy and rapid. Upon receiving a set of data, the data analyst must confer with the subject matter specialist(s) to make sure that he understands the entire problem. He must then "explore" the data, possibly in a number of ways. These cannot be specified in a flow chart, because of the immense variety of problems and different data sets that can be produced. This exploration is a creative process, and the data analyst may well have to confer with the subject matter specialist *throughout* the process. In the end,

the data analyst will have prepared charts and graphs to highlight what is important about the data, and to allow the subject matter specialist to draw his own conclusions. I do not believe that the statistician should make final decisions; he should, however, present those summaries and graphs that make it easy to "see the entire picture" and to make the appropriate inferences. His task is to draw attention to specific matters of interest; in general he cannot resolve the difficulties, since they are in the subject matter domain, not in that of the statistician.

13.4 DISCUSSION OF AN EXAMPLE

To illustrate in more detail the points made above, we look in some depth at a set of interlaboratory data. It was obtained using the Bekk instrument to measure the smoothness of 14 samples of paper. Fourteen laboratories participated, each making replicate measurements of smoothness on each of the 14 materials. However, two of the laboratories made measurements at a different relative humidity; they were eliminated from the analysis, since the relative humidity affects the results. The data in terms of averages and standard deviations of cells are shown in Tables 13.1 and 13.2. Figure 13.1 and 13.2 show the h and k plots, arranged in groups by laboratory. Figure 13.3 is a plot of repeatability and reproducibility standard deviation versus level of the measurement. These three graphs provide more useful information than sophisticated analytical methods would. Figure 13.1 shows us which cells are unusually large or unusually small; it also shows very clearly the pattern of the laboratory biases. Since the materials have been ordered in the original table from low to high, we see the decreasing trend in laboratories 1 and 2, the increasing trend in laboratories 4 and 10, the strong negative bias of laboratory 8, and a somewhat lesser negative bias of laboratories 11 and 12. We also see that laboratory 9 is generally positively biased, with a single sizeable exception. It is clear from this graph that the data definitely are nonadditive (trends in laboratories 1, 2, 4, 10). Figure 13.2 shows a fairly even pattern of k values, with possibly two exceptions, both in laboratory 3. Figure 13.3 shows a clear increase in both repeatability and reproducibility as the level increases. In both cases a linear approximation is seen to be sufficient. Table 13.3 shows us the values depicted in Figure 13.3.

Table 13.1 Bekk Smoothness of Paper: An Interlaboratory Study, Table of Cell Averages

Lab	B	J	C	D	I	L	M	E	A	G	H	F	K	N
1	6.38	6.75	12.41	14.43	14.44	18.58	42.00	45.70	86.80	110.40	154.20	143.70	164.80	191.40
2	5.60	6.38	13.06	14.90	15.20	18.14	41.50	44.60	88.70	102.70	154.20	160.10	170.60	198.20
3	5.25	5.35	11.95	13.70	13.43	15.10	37.90	43.60	78.70	114.90	137.30	151.20	178.10	173.50
4	4.01	5.88	11.73	13.41	12.70	16.16	40.60	40.70	91.30	100.40	167.00	207.10	207.00	201.00
5	4.36	4.67	10.53	11.55	13.79	14.59	32.90	41.20	78.40	99.40	129.90	179.30	173.70	173.60
6	4.12	5.25	9.62	11.63	14.25	15.38	36.50	37.50	81.90	99.80	150.90	161.00	166.40	182.50
7	4.50	5.88	11.25	12.63	13.00	15.38	35.60	40.50	80.60	112.40	155.30	165.60	186.30	205.60
8	3.75	4.38	9.75	11.25	11.25	13.75	31.00	31.90	65.10	90.10	126.00	139.80	154.80	162.30
9	4.45	6.16	13.01	13.75	15.09	17.01	35.00	44.10	90.10	105.30	148.10	187.00	198.70	210.90
10	4.42	5.59	12.75	13.35	14.66	17.08	43.00	47.50	92.00	115.10	172.40	201.50	213.60	217.70
11	3.98	4.75	9.92	11.70	11.25	14.74	35.60	37.20	79.00	91.90	131.80	150.10	171.20	186.20
12	3.55	4.29	9.25	11.56	12.50	15.10	37.90	37.60	75.80	95.80	129.20	149.40	172.80	174.90
Avg	4.53	5.44	11.27	12.82	13.46	15.92	37.46	41.09	82.37	103.18	146.36	166.32	179.83	189.82
SD	0.82	0.80	1.41	1.26	1.36	1.49	3.75	4.43	7.85	8.57	15.39	22.48	17.99	17.17

Table 13.2 Bekk Smoothness of Paper: An Interlaboratory Study, Table of Cell Standard Deviations

Lab	B	J	C	D	I	L	M	E	A	G	H	F	K	N
1	1.14	0.56	1.55	1.59	2.37	2.46	9.20	5.70	6.00	20.10	8.80	24.70	17.30	17.80
2	0.81	0.47	2.19	1.34	1.15	1.94	4.30	5.30	13.70	25.20	14.80	34.80	14.80	16.10
3	0.75	1.39	1.97	1.33	1.53	4.31	7.50	4.70	11.20	23.20	7.50	25.30	13.20	23.70
4	0.68	0.17	1.94	1.04	1.61	1.54	8.50	2.30	18.50	17.00	14.70	34.70	25.10	28.00
5	0.71	0.30	1.75	1.31	1.48	1.70	4.60	4.30	10.20	17.00	8.80	29.70	11.40	24.30
6	0.64	0.46	1.85	1.41	1.91	1.19	4.40	4.40	10.80	15.70	16.20	24.60	14.30	27.90
7	0.54	0.35	1.04	0.74	1.51	1.41	7.10	4.10	8.60	17.90	12.60	15.40	13.80	19.60
8	0.71	0.52	1.39	0.71	1.58	1.28	6.60	4.70	5.80	19.70	14.90	23.40	12.90	17.00
9	0.74	0.46	1.88	0.75	1.93	1.81	8.00	4.50	9.90	18.80	13.50	32.80	20.20	37.90
10	0.62	0.95	1.66	1.50	2.67	2.10	5.20	7.40	14.60	36.90	22.80	31.50	11.00	28.40
11	0.43	0.25	1.51	1.80	1.43	0.96	4.60	3.20	9.70	1.70	16.60	25.00	20.20	20.30
12	0.46	0.20	1.26	1.32	1.11	1.31	8.40	3.50	9.20	18.10	9.60	15.20	14.80	15.20
	POOLED STD. DEV. BETWEEN REPL., S_R (J)													
	0.71	0.61	1.70	1.28	1.75	2.02	6.76	4.67	11.22	20.73	14.02	27.17	16.25	23.89

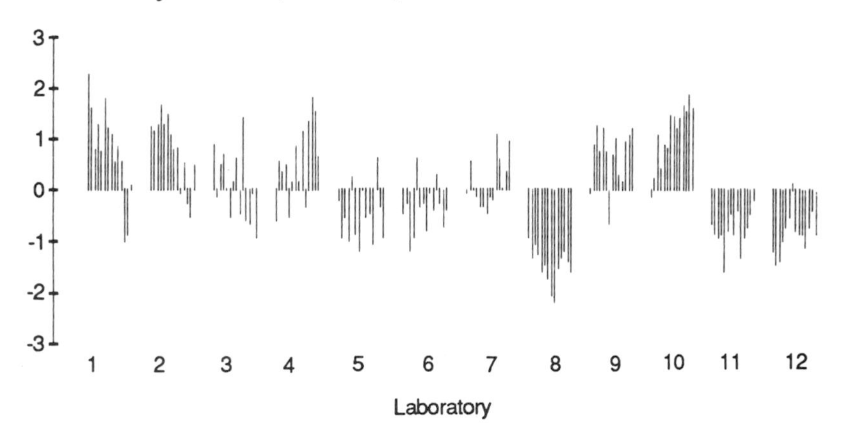

Figure 13.1 Bekk smoothness h-values by laboratories.

Since the data are definitely not additive, it may be of interest to explore what model does represent the data. Interlaboratory data very often follow a row-linear model (when the rows represent laboratories). A row-linear fit for out data is shown in Table 13.4. It exhibits the heights, slopes, and standard deviations of fit for the cell averages for each laboratory, when plotted against the column averages. The laboratories are ordered in increasing order of heights. When fitting straight lines or curves, we should always look at the residuals of the fit. They are shown in Figure 13.4, arranged in groups of materials. Clearly, they exhibit a marked increasing trend from materials with low Bekk value to materials with high Bekk value. Since each fitted line encompasses all materials, the fitting procedure violates the assumption, underlying classical straight line fitting, of equal error variance for all

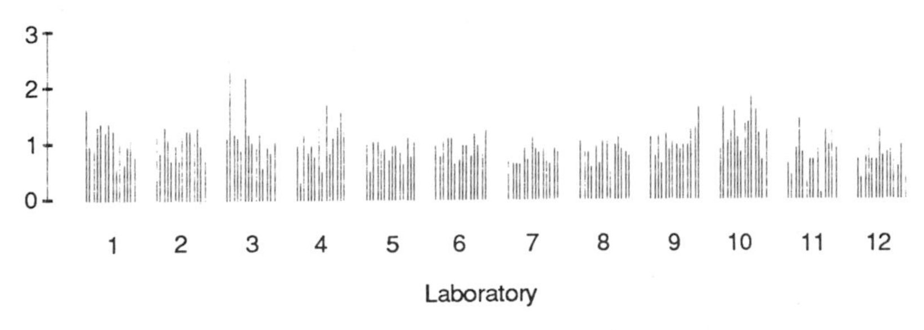

Figure 13.2 Bekk smoothness k-values by laboratories.

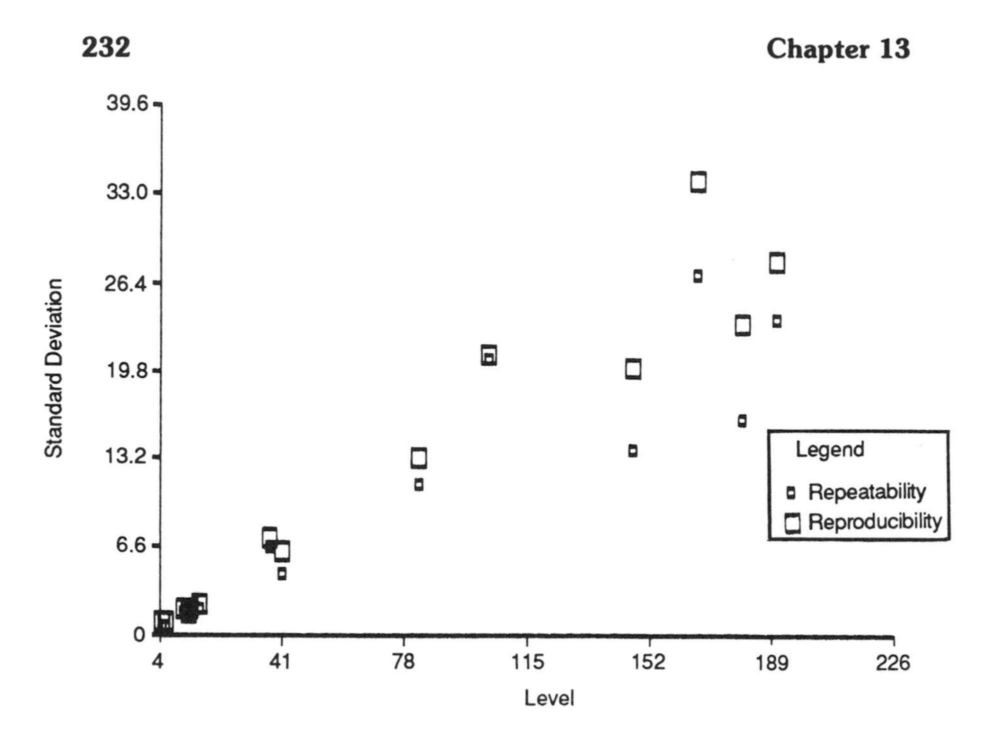

Figure 13.3 Bekk smoothness. Graph of Sr(J) & SR(J), standard deviations of repeatability and reproducibility versus level.

points. The appropriate remedy for this situation is a *weighted* procedure for the straight line fitting. Reasonable choices for the weights are the reciprocal of either the repeatability variance or the variances between cell averages for each material, but any choice has to be checked by plotting the equivalent of Figure 13.4, for residuals of the weighted analyses. In our case, the reciprocal of the variance between cell averages turns out to be a good choice. Figure 13.5 shows the residuals, in groups of materials, for this choice. Thus, a representation of our data is given by the model

$$y_{ij} = \alpha_1 + \beta_i x_j + \varepsilon_{ij}$$

where x_j is the column average, and ε_{ij} has a variance depending essentially only on j.

The standard deviation of ε_{ij} is equal to

$$s_{\varepsilon_{ij}} = -.02906 + .1041\, s_{B_j}$$

where s_{B_j} is the standard deviation between cell averages (over labo-

Table 13.3 Precision Parameters for Bekk Smoothness

Level	Avg	s_r (*J*)	s_R (*J*)
1	4.531	0.708	1.056
2	5.444	0.606	0.983
3	11.269	1.696	2.122
4	12.822	1.282	1.742
5	13.463	1.747	2.125
6	15.918	2.021	2.407
7	37.458	6.763	7.352
8	41.092	4.674	6.221
9	82.367	11.216	13.101
10	103.183	20.728	21.198
11	146.358	14.022	20.219
12	166.317	27.174	33.932
13	179.833	16.253	23.552
14	189.817	23.892	28.186

Table 13.4 Row-linear Fit for Bekk Smoothness Data

Lab	Height	Slope	Standard deviation
8	61.0807	0.8565	1.9613
12	66.4036	0.9220	3.0502
11	67.0957	0.9433	3.6852
5	69.1350	0.9646	6.9948
6	69.7679	0.9660	3.6409
3	69.9986	0.9384	5.9567
1	72.2850	0.9454	7.8297
2	73.8486	0.9878	4.9259
7	74.6100	1.0538	3.9537
9	77.7621	1.0966	4.6576
4	79.9993	1.1436	8.2964
10	83.6179	1.1818	3.1682

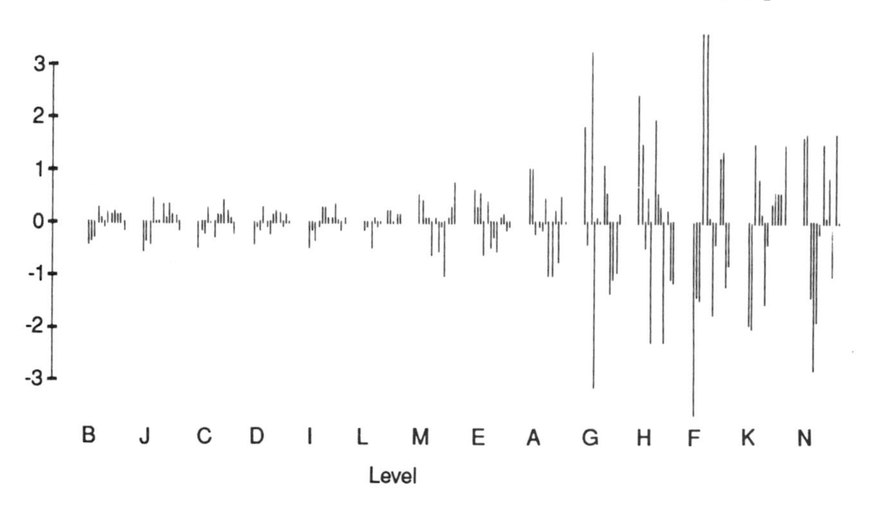

Figure 13.4 Bekk smoothness. Graph of standardized residuals by level.

ratories) for column *j*. The weight is the reciprocal of the square of this quantity.

It may be rewarding to find a model of this type, but in terms of our problem, the model is essentially superfluous. Figures 13.1, 13.2 and 13.3, together with Table 13.3 provided us with all the pertinent information.

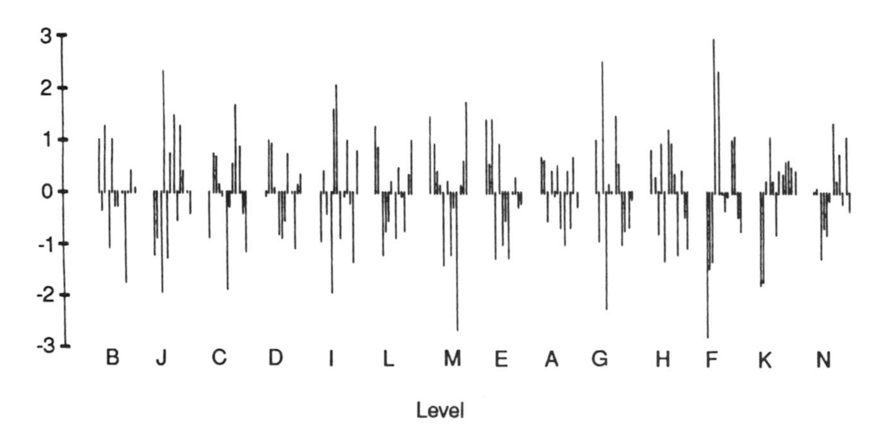

Figure 13.5 Bekk smoothness. Residuals by materials in weighted scale.

Our reason for discussing this example is to place it within the framework of our recommendations for data analysis.

The problem underlying this data set is well-stated. Our analysis is predominantly graphical. While each graph looks at a different aspect of the data, each graph covers the *entire* data set, not just a portion of it. The same holds for tables. We have not subjected the data to tests of significance, because statistical significance is mostly irrelevant in our case. We have pointed out potential outliers (the within-cell variability for cells 3J and 3L), but have *not* rejected them. Rejection of any data should be based predominantly on subject matter considerations.

We have found a reasonable model for the data but this model is essentially irrelevant for the purpose for which these data were obtained. It would certainly be possible to perform many statistical tests on these data, but they would mostly fall in the category of window dressing. The real purpose of data analysis is not to display an erudition in statistical tests, but rather to answer satisfactorily the practical questions that can be asked about the experiment.

REFERENCES

Cramer, Harold (1974). Mathematical Methods of Statistics. Princeton University Press, Princeton, New Jersey.

Fisher, R. A. (1955). Statistical Methods and Scientific Induction. J. Royal Statistical Society, *8*, 69–71.

Press, James (1989). Bayesian Statistics. Principles, Models, and Applications, John Wiley, New York.

Appendix

Table A1.a Critical Values of h and k at the 1.0% Level

		k Number of Replicates, n								
p	h	2	3	4	5	6	7	8	9	10
3	1.15	1.71	1.64	1.58	1.53	1.49	1.46	1.43	1.41	1.39
4	1.49	1.92	1.77	1.67	1.60	1.55	1.51	1.48	1.45	1.43
5	1.72	2.05	1.85	1.73	1.65	1.59	1.55	1.51	1.48	1.46
6	1.87	2.14	1.90	1.77	1.68	1.62	1.57	1.53	1.50	1.47
7	1.98	2.21	1.94	1.79	1.70	1.63	1.58	1.54	1.51	1.49
8	2.06	2.26	1.96	1.81	1.72	1.65	1.60	1.55	1.52	1.49
9	2.13	2.29	1.99	1.83	1.73	1.66	1.61	1.56	1.53	1.50
10	2.18	2.32	2.00	1.84	1.74	1.67	1.61	1.57	1.53	1.51
11	2.22	2.35	2.01	1.85	1.75	1.67	1.62	1.57	1.54	1.51
12	2.25	2.37	2.03	1.86	1.75	1.68	1.62	1.58	1.54	1.51
13	2.27	2.38	2.04	1.86	1.76	1.68	1.63	1.58	1.55	1.52
14	2.30	2.40	2.04	1.87	1.76	1.69	1.63	1.59	1.55	1.52
15	2.32	2.41	2.05	1.88	1.77	1.69	1.63	1.59	1.55	1.52
16	2.33	2.42	2.06	1.88	1.77	1.69	1.64	1.59	1.55	1.52
17	2.33	2.43	2.06	1.88	1.77	1.70	1.64	1.59	1.56	1.53
18	2.36	2.44	2.07	1.89	1.78	1.70	1.64	1.60	1.56	1.53
19	2.37	2.45	2.07	1.89	1.78	1.70	1.64	1.60	1.56	1.53
20	2.39	2.45	2.07	1.89	1.78	1.70	1.64	1.60	1.56	1.53
21	2.39	2.46	2.08	1.90	1.78	1.70	1.65	1.60	1.56	1.53
22	2.40	2.47	2.08	1.90	1.79	1.71	1.65	1.60	1.56	1.53
23	2.41	2.47	2.08	1.90	1.79	1.71	1.65	1.60	1.56	1.53
24	2.42	2.47	2.09	1.90	1.79	1.71	1.65	1.60	1.56	1.53
25	2.42	2.48	2.09	1.90	1.79	1.71	1.65	1.60	1.57	1.53
26	2.43	2.48	2.09	1.91	1.79	1.71	1.65	1.60	1.57	1.53
27	2.44	2.49	2.09	1.91	1.79	1.71	1.65	1.60	1.57	1.54
28	2.44	2.49	2.10	1.91	1.79	1.71	1.65	1.61	1.57	1.54
29	2.45	2.49	2.10	1.91	1.80	1.71	1.65	1.61	1.57	1.54
30	2.45	2.50	2.10	1.91	1.80	1.71	1.65	1.61	1.57	1.54

Table A1.b Critical Values of h and k at the 0.5% Level

		k Number of Replicates, n								
p	h	2	3	4	5	6	7	8	9	10
3	1.15	1.72	1.67	1.61	1.56	1.52	1.49	1.47	1.44	1.42
4	1.49	1.95	1.82	1.73	1.66	1.60	1.56	1.53	1.50	1.47
5	1.74	2.11	1.92	1.79	1.71	1.65	1.60	1.56	1.53	1.50
6	1.92	2.22	1.98	1.84	1.75	1.68	1.63	1.59	1.55	1.52
7	2.05	2.30	2.03	1.87	1.77	1.70	1.65	1.60	1.57	1.54
8	2.15	2.36	2.06	1.90	1.79	1.72	1.66	1.62	1.58	1.55
9	2.23	2.41	2.09	1.92	1.81	1.73	1.67	1.63	1.59	1.56
10	2.29	2.45	2.11	1.93	1.82	1.74	1.68	1.63	1.59	1.56
11	2.34	2.49	2.13	1.94	1.83	1.75	1.69	1.64	1.60	1.57
12	2.38	2.51	2.14	1.96	1.84	1.76	1.69	1.65	1.61	1.57
13	2.41	2.54	2.15	1.97	1.85	1.76	1.70	1.65	1.61	1.58
14	2.44	2.56	2.16	1.97	1.85	1.77	1.70	1.65	1.61	1.58
15	2.47	2.57	2.17	1.98	1.86	1.77	1.71	1.66	1.62	1.58
16	2.49	2.59	2.18	1.99	1.86	1.78	1.71	1.66	1.62	1.58
17	2.51	2.60	2.19	1.99	1.87	1.78	1.72	1.66	1.62	1.59
18	2.53	2.61	2.20	2.00	1.87	1.78	1.72	1.67	1.62	1.59
19	2.54	2.62	2.20	2.00	1.87	1.79	1.72	1.67	1.63	1.59
20	2.56	2.63	2.21	2.01	1.88	1.79	1.72	1.67	1.63	1.59
21	2.57	2.64	2.21	2.01	1.88	1.79	1.72	1.67	1.63	1.59
22	2.58	2.65	2.22	2.01	1.88	1.79	1.73	1.67	1.63	1.60
23	2.59	2.66	2.22	2.01	1.89	1.79	1.73	1.67	1.63	1.60
24	2.60	2.66	2.23	2.02	1.89	1.80	1.73	1.68	1.63	1.60
25	2.61	2.67	2.23	2.02	1.89	1.80	1.73	1.68	1.63	1.60
26	2.62	2.67	2.23	2.02	1.89	1.80	1.73	1.68	1.63	1.60
27	2.62	2.68	2.24	2.02	1.89	1.80	1.73	1.68	1.64	1.60
28	2.63	2.68	2.24	2.02	1.89	1.80	1.73	1.68	1.64	1.60
29	2.64	2.69	2.24	2.03	1.89	1.80	1.73	1.68	1.64	1.60
30	2.64	2.69	2.24	2.03	1.90	1.80	1.73	1.68	1.64	1.60

Table A1.c Critical Values of h and k at the 0.1% Level

		k Number of Replicates, *n*								
p	h	2	3	4	5	6	7	8	9	10
3	1.15	1.73	1.70	1.66	1.62	1.59	1.56	1.53	1.51	1.49
4	1.50	1.98	1.90	1.81	1.75	1.69	1.65	1.61	1.58	1.55
5	1.77	2.18	2.03	1.91	1.82	1.76	1.70	1.66	1.63	1.60
6	1.99	2.33	2.12	1.98	1.88	1.80	1.74	1.70	1.66	1.62
7	2.16	2.45	2.19	2.03	1.91	1.83	1.77	1.72	1.68	1.64
8	2.29	2.54	2.24	2.06	1.94	1.86	1.79	1.74	1.70	1.66
9	2.40	2.62	2.28	2.09	1.97	1.88	1.81	1.75	1.71	1.67
10	2.48	2.68	2.31	2.11	1.99	1.89	1.82	1.77	1.72	1.68
11	2.55	2.73	2.34	2.13	2.00	1.91	1.83	1.78	1.73	1.69
12	2.61	2.77	2.37	2.16	2.02	1.92	1.84	1.78	1.73	1.69
13	2.67	2.81	2.39	2.17	2.03	1.92	1.85	1.79	1.74	1.70
14	2.71	2.84	2.40	2.18	2.04	1.93	1.86	1.80	1.75	1.70
15	2.75	2.87	2.42	2.19	2.04	1.94	1.86	1.80	1.75	1.71
16	2.78	2.90	2.43	2.20	2.05	1.95	1.87	1.81	1.75	1.71
17	2.81	2.92	2.45	2.21	2.06	1.95	1.87	1.81	1.76	1.72
18	2.84	2.94	2.46	2.22	2.06	1.96	1.88	1.81	1.76	1.72
19	2.86	2.96	2.47	2.22	2.07	1.96	1.88	1.82	1.76	1.72
20	2.88	2.97	2.48	2.23	2.07	1.96	1.88	1.82	1.77	1.72
21	2.90	2.99	2.48	2.23	2.08	1.97	1.88	1.82	1.77	1.72
22	2.92	3.00	2.50	2.24	2.08	1.97	1.89	1.82	1.77	1.73
23	2.94	3.02	2.50	2.25	2.09	1.97	1.89	1.83	1.77	1.73
24	2.95	3.03	2.51	2.25	2.09	1.98	1.89	1.83	1.77	1.73
25	2.97	3.04	2.52	2.25	2.09	1.98	1.89	1.83	1.77	1.73
26	2.98	3.05	2.52	2.26	2.09	1.98	1.90	1.83	1.78	1.73
27	2.99	3.06	2.53	2.26	2.10	1.98	1.90	1.83	1.78	1.73
28	3.00	3.06	2.53	2.26	2.10	1.98	1.90	1.83	1.78	1.73
29	3.01	3.07	2.53	2.27	2.10	1.99	1.90	1.84	1.78	1.74
30	3.02	3.08	2.54	2.27	2.10	1.99	1.90	1.84	1.78	1.74

Table A2 Factors for Control Charts

Size of subgroup	Averages A_3	Standard deviations B_7	B_8
2	2.659	0.000	3.267
3	1.954	0.000	2.568
4	1.628	0.000	2.266
5	1.427	0.000	2.089
6	1.287	0.030	1.970
7	1.182	0.118	1.882
8	1.099	0.185	1.815
9	1.032	0.239	1.761
10	0.975	0.284	1.716
11	0.927	0.321	1.679
12	0.886	0.354	1.646
13	0.850	0.382	1.618
14	0.817	0.406	1.594
15	0.789	0.428	1.572
16	0.763	0.448	1.552
17	0.739	0.466	1.534
18	0.718	0.482	1.518
19	0.698	0.497	1.503
20	0.680	0.510	1.490
21	0.663	0.523	1.477
22	0.647	0.534	1.466
23	0.633	0.545	1.455
24	0.619	0.555	1.445
25	0.606	0.565	1.435

3-Sigma limits
- Averages
  - lower limits: $\bar{x} - A_3 \times \bar{s}$
  - upper limit: $\bar{x} + A_3 \times \bar{s}$
- Standard deviations
  - lower limit: $B_7 \times \bar{s}$
  - upper limit: $B_8 \times \bar{s}$

2-Sigma limits
- Averages
  - lower limit: $\bar{x} - \frac{2}{3} A_3 \times \bar{s}$
  - upper limit: $\bar{x} + \frac{2}{3} A_3 \times \bar{s}$
- Standard deviations
  - lower limit: $[1 - \frac{2}{3}(B_8 - 1)] \times \bar{s}$
  - upper limit: $[1 + \frac{2}{3}(B_8 - 1)] \times \bar{s}$

Table A.3 Areas of the Standard Normal Distribution

Z	0.00	0.01	0.02	0.03	0.04	0.05	0.06	0.07	0.08	0.09
0.0	0.0000	0.0040	0.0080	0.0120	0.0160	0.0199	0.0239	0.0279	0.0319	0.0359
0.1	.0398	.0438	.0478	.0517	.0557	.0596	.0636	.0675	.0714	.0753
0.2	.0793	.0832	.0871	.0910	.0948	.0987	.1026	.1064	.1103	.1141
0.3	.1179	.1217	.1255	.1293	.1331	.1368	.1406	.1443	.1480	.1517
0.4	.1554	.1591	.1628	.1664	.1700	.1736	.1772	.1808	.1844	.1879
0.5	.1915	.1950	.1985	.2019	.2054	.2088	.2123	.2157	.2190	.2224
0.6	.2257	.2291	.2324	.2357	.2389	.2422	.2454	.2486	.2517	.2549
0.7	.2580	.2611	.2642	.2673	.2704	.2734	.2764	.2794	.2823	.2852
0.8	.2881	.2910	.2939	.2967	.2995	.3023	.3051	.3078	.3106	.3133
0.9	.3159	.3186	.3212	.3238	.3264	.3289	.3315	.3340	.3365	.3389
1.0	.3413	.3438	.3461	.3485	.3508	.3531	.3554	.3577	.3599	.3621
1.1	.3643	.3665	.3686	.3708	.3729	.3749	.3770	.3790	.3810	.3830
1.2	.3849	.3869	.3888	.3907	.3925	.3944	.3962	.3980	.3997	.4015
1.3	.4032	.4049	.4066	.4082	.4099	.4115	.4131	.4147	.4162	.4177
1.4	.4192	.4207	.4222	.4236	.4251	.4265	.4279	.4292	.4306	.4319
1.5	.4332	.4345	.4357	.4370	.4382	.4394	.4406	.4418	.4429	.4441

Table A.3 (continued)

Z	0.00	0.01	0.02	0.03	0.04	0.05	0.06	0.07	0.08	0.09
1.6	.4452	.4463	.4474	.4484	.4495	.4505	.4515	.4525	.4535	.4545
1.7	.4554	.4564	.4573	.4582	.4591	.4599	.4608	.4616	.4625	.4633
1.8	.4641	.4649	.4656	.4664	.4671	.4678	.4686	.4693	.4699	.4706
1.9	.4713	.4719	.4726	.4732	.4738	.4744	.4750	.4756	.4761	.4767
2.0	.4772	.4778	.4783	.4788	.4793	.4798	.4803	.4808	.4812	.4817
2.1	.4821	.4826	.4830	.4834	.4838	.4842	.4846	.4850	.4854	.4857
2.2	.4861	.4864	.4868	.4871	.4875	.4878	.4881	.4884	.4887	.4890
2.3	.4893	.4896	.4898	.4901	.4904	.4906	.4909	.4911	.4913	.4916
2.4	.4918	.4920	.4922	.4925	.4927	.4929	.4931	.4932	.4934	.4936
2.5	.4938	.4940	.4941	.4943	.4945	.4946	.4948	.4949	.4951	.4952
2.6	.4953	.4955	.4956	.4957	.4959	.4960	.4961	.4962	.4963	.4964
2.7	.4965	.4966	.4967	.4968	.4969	.4970	.4971	.4972	.4973	.4974
2.8	.4974	.4975	.4976	.4977	.4977	.4978	.4979	.4979	.4980	.4981
2.9	.4981	.4982	.4982	.4983	.4984	.4984	.4985	.4985	.4986	.4986
3.0	.4987	.4987	.4987	.4988	.4988	.4989	.4989	.4989	.4990	.4990
3.1	.4990	.4991	.4991	.4991	.4992	.4992	.4992	.4992	.4993	.4993
3.2	.4993	.4993	.4994	.4994	.4994	.4994	.4994	.4995	.4995	.4995
3.3	.4995	.4995	.4995	.4996	.4996	.4996	.4996	.4996	.4996	.4997
3.4	.4997	.4997	.4997	.4997	.4997	.4997	.4997	.4997	.4997	.4998
3.6	.4998	.4998	.4999	.4999	.4999	.4999	.4999	.4999	.4999	.4999
3.9	.5000									

Table A4 Degress of Freedom Corresponding to the First Three Eigenvalues for an $r \times s$ Table*

k	r/s	2	3	4	5	6	7	9	11	15	19
1	2	3.55	5.03	6.34	7.53	8.96	10.15	12.77	15.09	19.58	24.10
	3	5.03	6.71	8.33	9.85	11.32	12.74	15.56	18.08	23.23	28.34
	4	6.34	8.33	10.21	11.81	13.40	15.03	18.16	21.00	26.51	31.83
	5	7.53	9.85	11.81	13.51	15.47	16.97	20.23	23.24	29.14	34.90
	6	8.96	11.32	13.40	15.47	17.17	19.11	22.40	25.63	31.56	37.58
	7	10.15	12.74	15.03	16.97	19.11	20.96	24.35	27.77	34.24	40.30
	9	12.77	15.56	18.16	20.23	22.34	24.35	27.99	32.04	38.79	45.38
	11	15.09	18.08	21.00	23.24	25.63	27.77	32.04	36.25	42.73	50.33
	15	19.58	23.23	26.51	29.14	31.56	34.24	38.79	42.73	50.84	58.76
	19	24.10	28.34	31.83	34.90	37.58	40.30	45.38	50.33	58.76	65.94
	31	37.88	42.87	47.26	50.70	54.71	57.34	63.52	68.65	78.89	88.15
	49	57.85	63.78	69.31	73.46	77.51	81.39	88.82	95.34	106.82	117.83
	99	111.22	120.09	127.27	133.44	138.99	144.25	152.80	161.46	176.77	192.05
2	2	0.45	0.97	1.66	2.47	3.04	3.85	5.23	6.91	10.42	13.90
	3	0.97	2.01	3.01	4.00	4.98	6.01	7.98	10.02	14.04	17.96
	4	1.66	3.01	4.21	5.45	6.66	7.88	10.12	12.50	16.90	21.25
	5	2.47	4.00	5.45	7.02	8.22	9.70	12.08	14.77	19.62	24.28
	6	3.04	4.98	6.66	8.22	9.76	11.28	14.06	16.79	22.05	27.10
	7	3.85	6.01	7.88	9.70	11.28	12.82	15.78	18.76	24.14	29.59

Table A4 (continued)

k	r/s	2	3	4	5	6	7	9	11	15	19
	9	5.23	7.98	10.12	12.08	14.06	15.78	19.30	22.18	28.59	34.00
	11	6.91	10.02	12.50	14.77	16.79	18.76	22.18	25.60	32.42	38.45
	15	10.42	14.04	16.90	19.62	22.05	24.14	28.59	32.42	39.55	46.40
	19	13.90	17.96	21.25	24.28	27.10	29.59	34.00	38.45	46.40	53.54
	31	24.12	29.99	34.26	38.32	41.60	44.55	50.38	55.57	65.47	74.01
	49	40.15	48.10	53.79	58.20	62.55	66.40	73.25	80.24	90.97	101.74
	99	86.78	97.81	106.11	112.48	118.85	123.83	133.46	142.55	157.22	171.24
3	3		0.28	.66	1.15	1.69	2.25	3.46	4.89	7.72	10.70
	4		0.66	1.37	2.22	3.06	3.76	5.43	7.11	10.70	14.33
	5		1.15	2.22	3.19	4.16	5.24	7.24	9.22	13.24	17.17
	6		1.69	3.06	4.16	5.44	6.51	8.85	11.06	15.47	19.73
	7		2.25	3.76	5.24	6.51	7.81	10.44	12.82	17.55	22.24
	9		3.46	5.43	7.24	8.85	10.44	13.43	11.07	21.32	26.37
	11		4.89	7.11	9.22	11.06	12.82	16.07	19.07	25.02	30.58
	15		7.72	10.70	13.24	15.47	17.55	21.32	25.02	31.71	37.95
	19		10.70	14.33	17.17	19.73	22.24	26.37	30.58	37.95	44.89
	31		20.14	25.27	29.30	32.66	35.60	41.34	46.33	55.94	64.20
	49		35.12	41.77	47.46	51.66	55.60	62.55	68.84	79.94	90.41
	99		79.09	89.71	97.33	103.25	109.53	119.41	127.69	143.41	156.36

*If the SVD is carried out on residuals from an additive fit. Then use $r - 1$ and $s - 1$.

Index